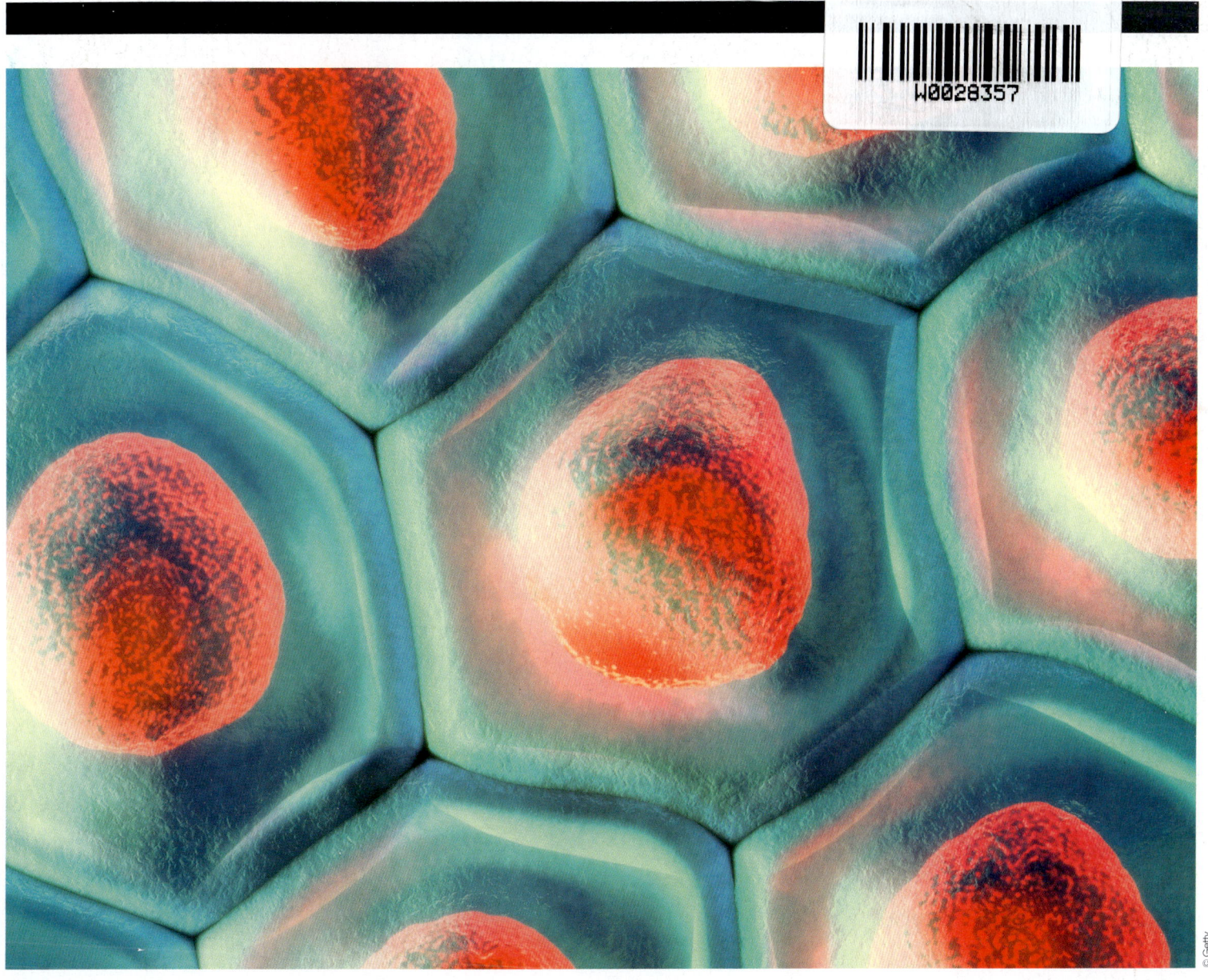

Welcome

Biology is the branch of science that explores all living organisms and the processes that create and govern them. It studies the animal kingdom, humans, plant life (including fungi and moulds) and the tiny world of microbiology, including micro-organisms like bacteria and viruses, and the cells that make up the building blocks of all living things. In this book, from the makers of *How It Works*, you'll discover more about these branches of biology and some of their key processes, like evolution, hormones, germ theory and photosynthesis, along with many other fascinating aspects of the study of life.

THE SCIENCE COLLECTION

Contents

THE ANIMAL KINGDOM

08 How life evolves
Biological pathways through time

14 Meet the mutants
How changes in DNA can lead to abnormalities and development

22 Animal domestication
How we created farm animals

26 Why animals shed
Why some creatures lose layers

HUMAN BIOLOGY

34 Journey through your blood vessels
Under the surface of the skin

40 Unravelling the nervous system
Explore this electric network and how it sends signals around your body

48 How hormones control your body
Endocrine signalling explained

56 Your immune system explained
Meet the cells and organs that keep deadly invaders at bay

64 How babies are made
The fusion of sex cells

08

MICROBIOLOGY

72 The power of pasteurisation
Keeping food safe from germs

76 How bleach kills germs
How a common household cleaner destroys bacteria and viruses

78 Lab-grown meat explained
How scientists are developing lab-grown meat from cell cultures

80 What is nanotech?
Enter an invisible world of tiny machines and medicines

88 What are microplastics?
What are these miniscule pollutants and why are they a problem?

94 What makes things biodegradable?
The biochemistry behind how things break down

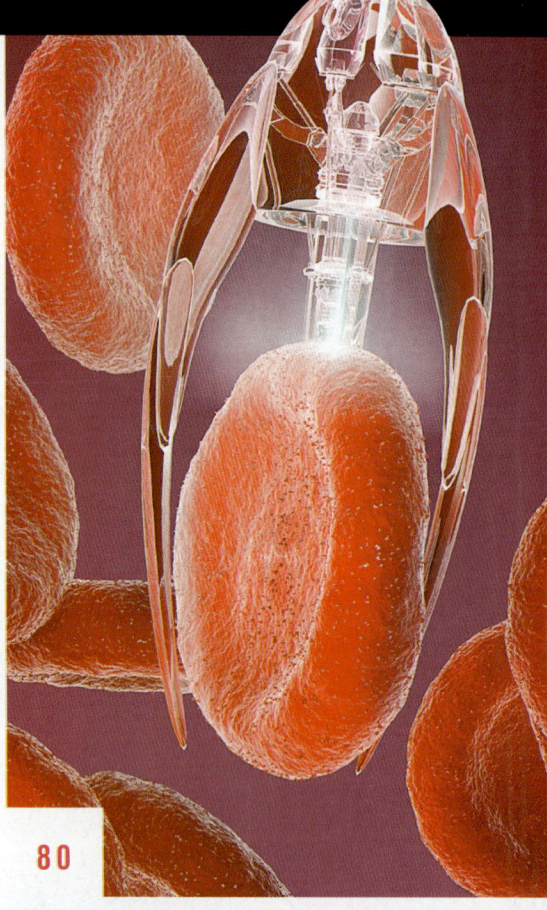
80

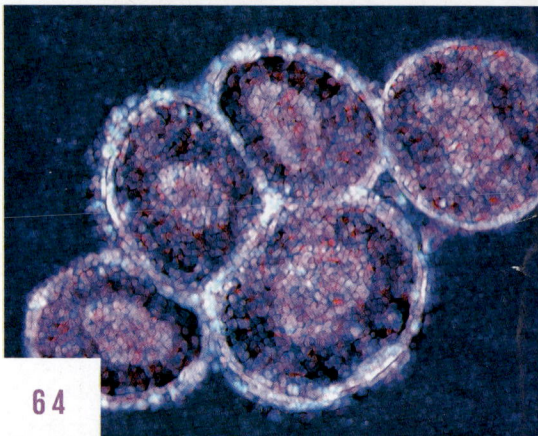
64

126

CONTENTS

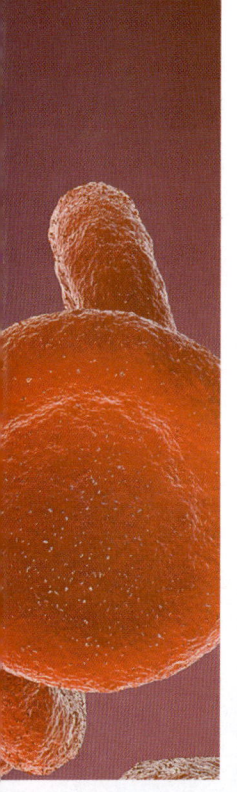
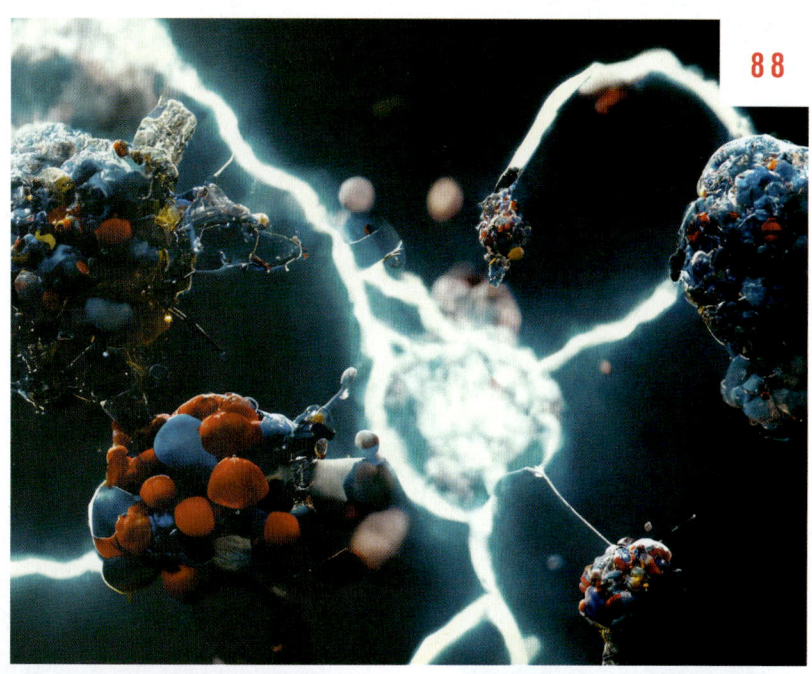

88

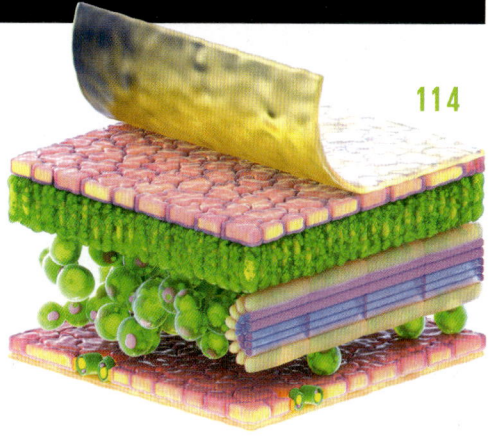

114

PLANTS & FUNGI

102 **The oldest plants in the world**
Ancient organisms

108 **Fascinating fruit**
Vehicles for seeds

114 **Why leaves turn brown in autumn**
Deciduous plants

116 **Why is grass green?**
Molecular colour

117 **Why nettles sting**
The science behind the sting

118 **What makes chilli peppers spicy?**
Tasty chemical compounds

120 **How to eat poisonous plants**
Removing toxic compounds

122 **Toxic mushrooms**
Find out why some kinds of fungi are deadly, and how to spot them

126 **The weird world of mould**
Find out more about this strange fungi-like group of organisms

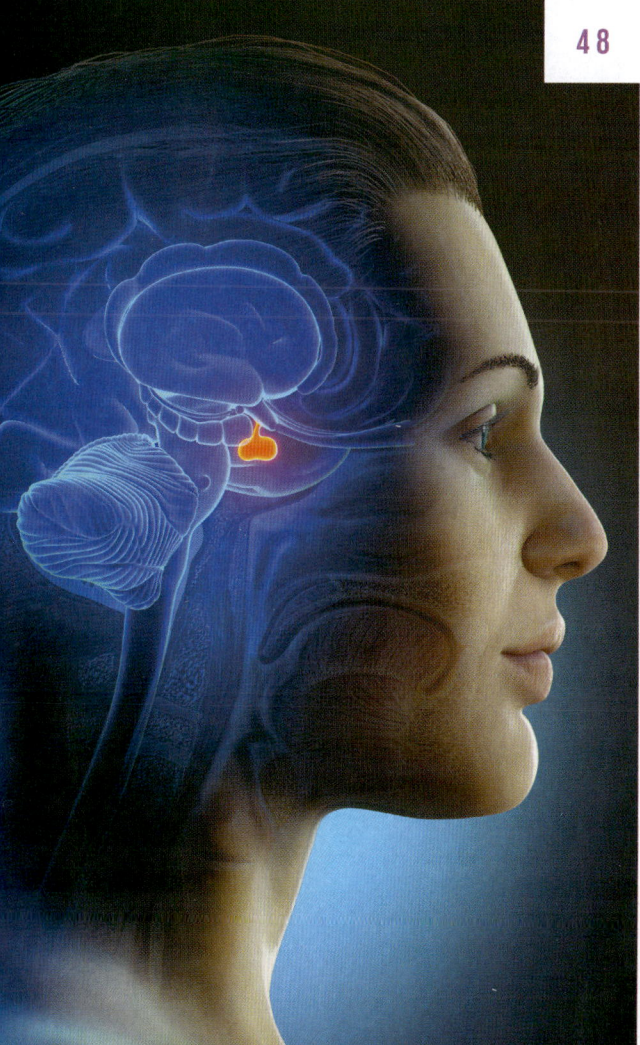

48

26

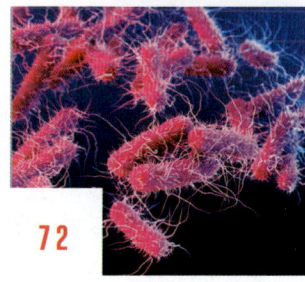

72

108

5

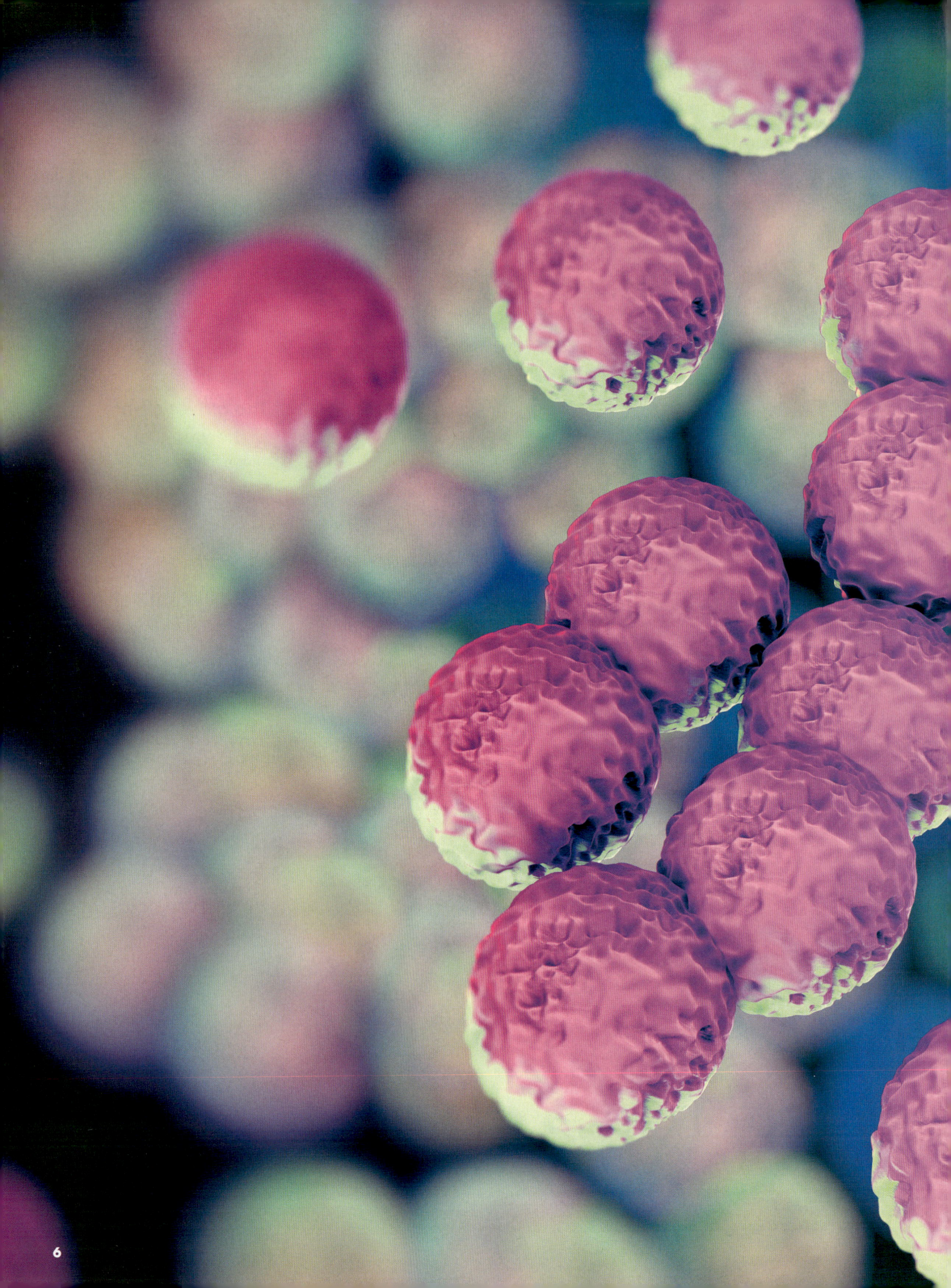

The Animal Kingdom

UNDERSTANDING BIOLOGY

How life evolves

Discover the biological pathways through time that led to the plants and animals of today

THE ANIMAL KINGDOM

HOW LIFE EVOLVES

Did you know?
The first reptiles evolved around 315 million years ago

For around 3.7 billion years, life has followed a roadmap of evolution. The journey to modern-day life and the more than 2.16 million known species that reside on Earth has been guided by several different evolutionary pathways. The naturalist known as 'the father of evolution', Charles Darwin, first proposed the theory of evolution by natural selection after he jumped aboard the HMS Beagle in 1831 and sailed around the world for five years. On his travels, he discovered diverse wildlife, observing that they had anatomy specially adapted to their environments. He concluded that a species' survival depends on traits that suit its environment. This survival-of-the-fittest approach means that only animals with beneficial characteristics can continue to reproduce and pass them onto their offspring. Darwin proposed that through forces of evolution, an ancestral family tree formed below modern-day species, leading to a common ancestor at its roots.

Forces of evolution are varied and specific to an ecology niche. For many, evolution is driven by sexual selection, whereby qualities in potential mates – such as plumage vibrancy or big antlers – are chosen by a partner to successfully reproduce. For others, natural selection occurs when an aspect of a species' environment is altered, such as by climatic change, and they are forced to adapt or die. Those that can change go on to reproduce, leading to some anatomical and behavioural differences over millennia, creating new species along the way. Today scientists refer to the creation of a new species as divergent evolution, where a new species separates from a common ancestor. Building on the work of Darwin, the term 'divergent evolution' was first introduced by evolutionist John Thomas Gulick in 1890. As the name suggests, divergent evolution occurs when a subgroup within a species diverges anatomically from the rest of the species, with these genes spreading until a new distinct species emerges. All life on Earth has undergone divergent evolution from a distant common ancestor, branching off into new families that form the tree of life.

For example, humans share more than 97 per cent of their DNA with the other great ape species on Earth: orangutans, gorillas and chimpanzees. Evolving from a common

9

UNDERSTANDING BIOLOGY

ancestor more than 25 million years ago, the great apes have diverged from their common lineage to produce an unknown number of now-extinct species that have paved the way for modern-day animals. Although they share a common ancestor, each ape has evolved distinctly different characteristics that have benefited them in their environment but have ultimately led to development of each individual ape species.

Divergent evolution occurs over long periods, with some studies suggesting that it generally takes around 2 million years for a new species to emerge. However, not all animals are willing to wait that long, instead rapidly evolving several new species through a process known as adaptive radiation. One of Darwin's most notable observations involved a group of finches that lived throughout the Galápagos Islands. Across the different islands, Darwin observed that among the related finches, beak size and shape varied depending on what food they ate. He concluded that with each new generation, those that were better equipped to find food were able to survive – a concept that formed the basis for his famous theory on natural selection.

Similarly, in Hawaii, a group of birds called honeycreepers have undergone adaptive radiation to produce a diverse family of birds with beaks specialised for their food. Some boast big, long beaks to probe underneath bark to find insects, while others developed stout robust beaks for cracking seeds and picking berries. Upon reaching Hawaii around 3 million years ago, the common ancestor of the honeycreeper species wasted no time adapting to the array of food sources found across the archipelago. Within 2 million years, 50 species of honeycreeper emerged. But only 17 species remain today.

Where divergent evolution leads to the creation of entirely new species from a common ancestor, another evolutionary pathway, known as convergent evolution, produces new species that have evolved surprisingly similar traits. At first glance you might forgive early biologists for mistaking dolphins for fish, or bats for a member of the bird family. That's because they share similar characteristics: dolphins evolved fish-like fins to swim, and bats developed bird-like wings for flight. However, dolphins and fish do not share a common ancestor, and neither do birds and bats. The two have completely separate evolutionary lineages, yet both

Birds and bats both evolved separately, resulting in wings despite having different lineages

Honey badgers evolved thick skin to evade the sting of a bee, but bees co-evolved a signalling system to attack badgers en masse

Human family tree

The road to becoming one of the great apes of the world

1.8 million years ago
The formation of the Congo River formed a barrier, splitting chimpanzees into two groups, one of which evolved into the modern-day bonobo.

6 million years ago
The evolutionary path of humans diverged from a primate ancestor shared with one of our closest living relatives today, the chimpanzee.

6 to 8 million years ago
Gorillas diverged from hominids. As more species of apes became extinct over time, gorillas evolved to become the largest of the great apes.

12 to 16 million years ago
Orangutans diverged from other apes of the Hominidae family in Asia, while the ancestors of the other great apes stayed in Africa.

Common ancestor
Though an exact common ancestor is unknown, a primate in the family Hominidae is responsible for starting the journey of ape evolution.

THE ANIMAL KINGDOM

developed similar anatomical features, known as analogies. When two species occupy the same ecological niche, and are therefore subject to the same evolutionary pressures, both evolve and adapt in similar ways, over time changing into new species that bear a resemblance to one another. For birds, the ability to fly emerged around 160 million years ago when a long-feathered dinosaur called Archaeopteryx, flapped its long feathers – which covered the skin of dinosaurs for millions of years. The elongated length of this dinosaur's arms helped generate some lift beneath them, allowing it to fly for short periods of time, in a similar way to a modern-day pheasant. Flight gave Archaeopteryx an advantage over other animals when evading predators, and also hunting prey from treetops. Bats, on the other hand, evolved only 50 million years ago, descending from an unknown mammal species. Scientists think that bats evolved from small tree-dwelling mammals that eventually developed flaps of skin between their toes called interdigital webbing. This eventually evolved into wings that not only allowed them to glide from tree to tree, but actively generate lift like a bird and fly.

Beneath the umbrella of divergent evolution there are microevolutions that can lead plants and animals down an evolutionary path that intertwines with another. Known as coevolution, this phenomenon occurs when two species develop anatomical characteristics as a result of interactions with another species. These interactions typically centre around one of three processes in nature: pollination, parasitism and predation.

The Honeycreepers
The birds with beaks tailored to their diets

'Apapane
(Himatione sanguinea)
These narrow-beaked birds feed on plant nectar, predominantly from the flowers of the 'ohi'a lehua tree.

Kaua'i nukupu'u
(Hemignathus hanapepe)
The thin hooked beaks of these birds are used to pick out insect prey that are found beneath tree bark.

'I'iwi
(Vestiaria coccinea)
To reach the nectar of tubular flowers, these birds have evolved long curved beaks.

Maui parrotbill
(Pseudonestor xanthophrys)
Also known as a kiwikiu, these birds use their parrot-like beaks to split branches to extract insects and their larvae.

Laysan finch
(Telespiza cantans)
As omnivores, their beaks are used to pick fruit and seeds, as well as pick off invertebrates and even eat carrion.

Lesser 'akialoa
(Akialoa obscura)
This extinct species used its beak to hunt insects living in tree bark, inserting its entire bill into crevices.

One of Darwin's large ground finches cracking a seed

A swallowtail caterpillar looking remarkably like a green snake

Did you know? There are 42 dolphin species in the world

UNDERSTANDING BIOLOGY

Pollinators such as bees and birds have evolved side by side with plant species in ways that benefit them both. For example, the pollen and nectar of the Darwin orchid (Angraecum sesquipedale) are found at the end of a long tube past its petals, which few pollinators could reach. Luckily, one pollinator, Wallace's sphinx moth (Xanthopan morganii praedicta), has evolved alongside the orchid and developed an enormous tongue called a proboscis. This can extend up to almost 30 centimetres for lapping up Darwin orchid nectar, picking up pollen along the way. It's not just the orchid that benefits – this also provides the sphinx moth with a food source other pollinators can't access.

Unlike coevolution that benefits both species, as seen in plant-and-pollinator relationships, some animals, such as Alcon blue caterpillars (Phengaris alcon), have co-evolved with other species in ways that make them the perfect parasites. By mimicking the scent emitted by the larvae of Myrmica ants, Alcon blue caterpillars can infiltrate a nest and be welcomed with open arms by the ant colony, which will then care for them as their own. Studies have also shown that when a nest is under threat, the emitted scent causes ants to prioritise the protection of the caterpillars over their own larvae.

Coevolution between predators and prey provides some of the most perplexing examples of nature's ability to adapt to an environment. Typically known as Batesian mimicry, it's named after the 19th-century naturalist Henry Walter Bates and occurs when a species has evolved, behaviourally or anatomically, to mimic the characteristics of another more dangerous or venomous animal. For example, the caterpillar of the spicebush swallowtail butterfly (Papilio troilus) has evolved to present markings that bear strong resemblance to the head of a smooth green snake (Opheodrys vernalis).

The purpose of displaying such a visual deception is to fool potential predators into believing that they are not the prey the predator might have hoped for, and are instead a dangerous animal that they don't want to interact with. The appearance of the spicebush swallowtail caterpillar wouldn't have the same effect without the existence of the smooth green snake, and thus the result of the snake's evolution has had a direct impact on the evolution of the spicebush swallowtail, as it needs to be a convincing enough mimic to deter predators.

Did you know? Animal pollination evolved 99.6 to 65.5 million years ago

Finding their fins
How convergent evolution created similar species from different ancestors

1 Unknown reptile
The evolutionary ancestors of prehistoric marine reptiles descended from land-dwelling reptiles that lived more than 300 million years ago.

2 Pakicetus
Around 50 million years ago, the ancestors of modern-day dolphins separated from their terrestrial ancestors.

3 Ichthyosaur
First appearing around 250 million years ago, this now-extinct animal was a fish-like reptile that scientists estimate could swim at up to 25 miles per hour.

4 Dolphin
Around 11 million years ago, the dolphin family Delphinidae first appeared.

5 Similar species
Although ichthyosaurs and dolphins are not related and don't share a common ancestor, all three share characteristics adapted for life in the ocean, such as fins.

THE ANIMAL KINGDOM

HOW LIFE EVOLVES

Surprising coevolution
Amazing examples of how two species have steered each other's evolution

The large eyes of a slow loris resemble the markings on a cobra's head

Slow loris and cobra
These furry tree climbers are surprisingly similar to snakes and have evolved to mimic some of their notable traits. The slow loris is one of the only known mammal species to produce venom for defence, which it secretes from a gland in its elbow and then licks to create a venomous saliva in its mouth that coats its teeth. When threatened, the slow loris hisses and will cross its arms over its head, making its face, complete with large eyes, resemble the shape and markings of a cobra.

A scoliid wasp mistaking a mirror orchid for a female wasp

Mirror orchid and scoliid wasp
In the interest of pollination, the mirror orchid (Ophrys speculum) has evolved an unusual petal that attracts male scoliid wasps (Dasyscolia ciliata). The petal, known as a labellum, closely resembles the furry brown bodies of female scoliid wasps. Tricked into thinking he's found a mate, the male wasp lands on the flower's imitation in hope of reproducing. Instead the male wasp finds itself with a head covered in pollen as it rubs against the plant's anther, an organ that contains pollen. The wasp will then pollinate another mirror orchid when it once again mistakes its petals for another potential mate.

Fig wasps emerging from their eggs inside a fig

Wasps inside figs
Figs produced by the fig tree (Ficus carica) are known as an inflorescence, not a fruit, where the inside of its bulbous body is full of clusters of small flowers and seeds. With its flowers on the inside of the fig, its only pollinator, a queen fig wasp (Eupristina verticillata), burrows into the fig's outer flesh and lays her eggs within before ultimately dying and being digested by the fig for nourishment. When the wasp's offspring hatch, the females gather pollen from inside the fig and take it to the next fig, where they will eventually lay their own eggs.

Cuckoo eggs are slightly larger than those in a warbler's brood

Cuckoo egg imposters
The common cuckoo (Cuculus canorus) is one of the worst parents in nature – mainly because as soon as the opportunity presents itself, female cuckoos will ditch their eggs in the nests of other birds for them to raise as their own. The presence of the Trojan egg is unknown to the host parent because cuckoos have evolved an egg colouration and speckled appearance that resembles the eggs of many other bird species, such as warblers, wagtails and pipits. The new host parent will raise the new addition until it's ready to fly the nest and continue the parasitic cycle.

UNDERSTANDING BIOLOGY

Meet the mutants

How changes in DNA can cause unusual abnormalities, but also shape the evolution of a species

THE ANIMAL KINGDOM

MEET THE MUTANTS

What makes you who you are? It's a molecule called deoxyribonucleic acid, better known as DNA. This omnipotent molecule is made up of twisted strands of sugars and phosphates, creating its double-helix shape. Connecting the twisting strands are four bases known as nucleotides: adenine, cytosine, guanine and thymine. The sequence in which these bases find themselves along each strand of DNA is what ultimately determines the development and function of an organism.

The way DNA works is similar to the way a computer reads binary code. A molecule of DNA is divided by varying lengths of base sequences called genes. Some genes contain hundreds of bases, others millions. Each gene acts as a set of instructions to program an organism, controlling what it looks like, how its body functions and even how it behaves. The complete set of genetic information in an organism is known as its genome. The human genome is held in 23 pairs of chromosomes, long bundles of DNA which sit inside the nucleus of a cell. The size of an organism's genome varies across the spectrum of life – for example, a fruit fly has only 4 chromosome pairs, while a dog has 39.

There are two important distinctions within a DNA molecule: genes that are referred to as coding genes and non-coding genes. Around one per cent of DNA is made up of coding genes, which actively supply the information for the production of proteins. Proteins are vital molecules that act as the raw ingredients to grow and develop an organism. The remaining 99 per cent of DNA is made of non-coding genes; these don't provide the blueprints for protein production but can influence the way in which proteins are made and even prevent production.

Like lines of computer code, when nucleotide sequences within a gene are changed, deleted or swapped, an organism's biological program is altered. This is known as a mutation. Mutations can be split into one of three categories: silent, missense and nonsense. Silent mutations are, as the name suggests, an alteration of nucleotides that have no protein production. Missense mutations, on the other hand, will alter the function of a protein. Finally, nonsense mutations cause proteins to be nonfunctional. Mutations occur during the time the cells of an organism divide and grow. During cell division, all the DNA information within the nucleus is unzipped and exactly copied to create a perfectly functioning replica. However, on some occasions the genetic duplication isn't exact –

Did you know? Down's syndrome is caused by a chromosome mutation

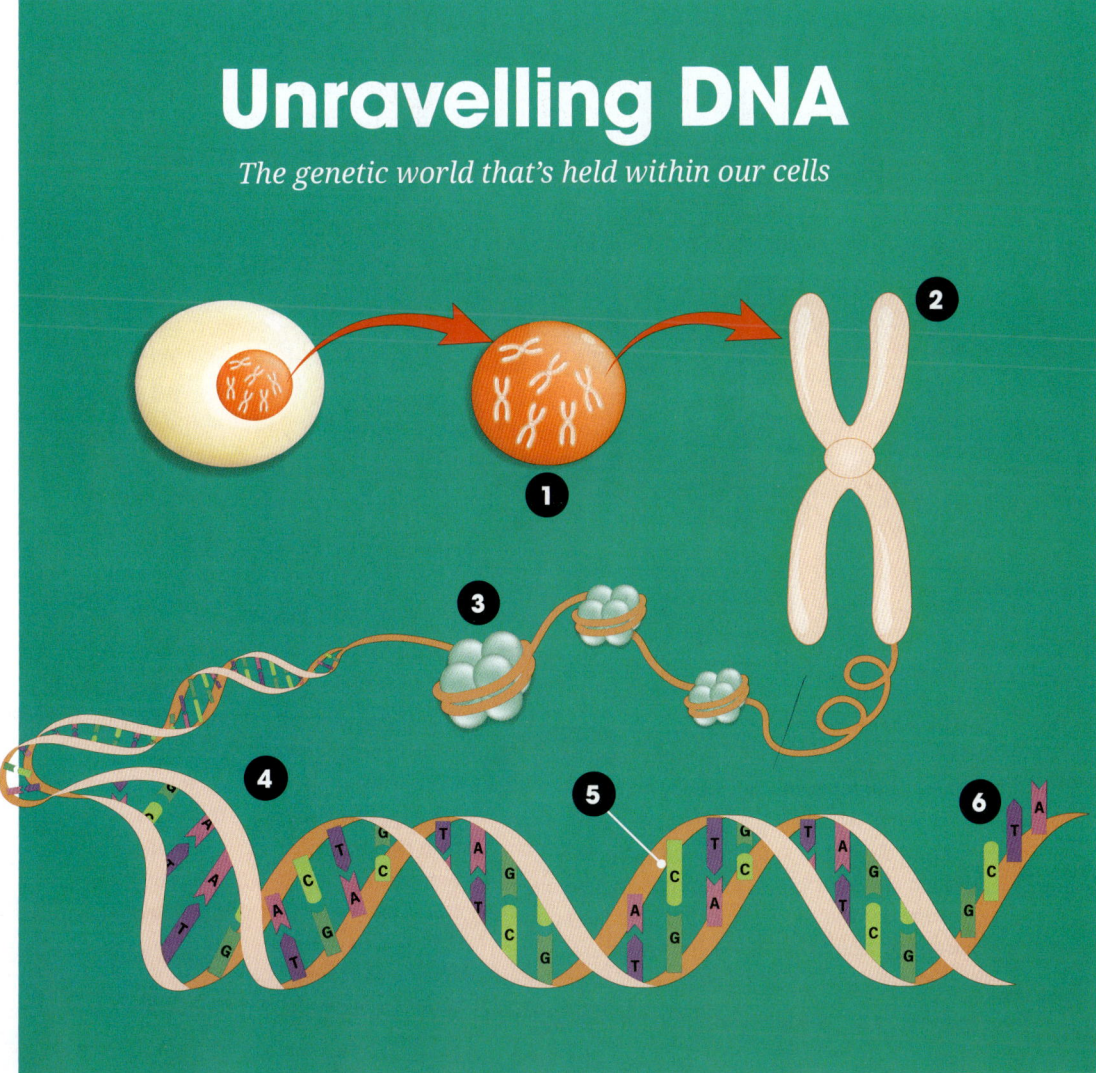

Unravelling DNA
The genetic world that's held within our cells

1 Nucleus
This is the location in the cell where all an organism's genetic information is held.

2 Chromosomes
Within each cell nucleus are 23 bundles of genetic information that are called chromosomes.

3 Histones
These basic proteins support the structure of a chromosome.

4 DNA
The structure of DNA is made up of a twisting double helix with a sugar phosphate backbone.

5 Nucleotides
Within DNA there is one of four nucleotides, also known as bases. These structural building blocks of DNA are teffectively he blueprints of an organism.

6 Coding
Nucleotides form specific pairs – adenine with thymine and cytosine with guanine. The arrangements in nucleotide sequences code for the production of certain proteins.

UNDERSTANDING BIOLOGY

some information is left behind or replaced, and a mutation occurs.

The effects of a mutation are wide-ranging. Tiny-point mutations, which occur at a single nucleotide base pair, occur trillions of times among the trillions of cells that make up an organism's body. The majority of these will be silent mutations that have no notable effect on the organism. However, there are occasions when these mutations can have implications to the physical appearance of an organism, how it behaves or cause a myriad of health issues. As an example, polymelia is an uncommon condition caused by a genetic mutation that leads to the development of additional limbs. The condition is rarely seen in humans but is occasionally reported in domestic animals such as chickens and sheep.

As to why mutations occur, scientists still don't have a complete picture. There isn't much scientific rhyme or reason when it comes to spontaneous mutations. Sometimes, during the process of cell division, errors are made in the replication of DNA, which results in a mutation. There are also active agents in the world that can tip the scales of genetic probability and cause a mutation. Active agents that cause a mutation are called mutagens and are either chemical or radioactive. Chemical mutagens are toxins that can alter DNA and lead to a whole host of health issues and congenital abnormalities. For example, the inhalation of asbestos can cause mutations that lead to the development of lung cancer.

One of the most striking physical mutations seen in the animal kingdom is known as bicephaly, or axial bifurcation, whereby an organism develops a two-headed but otherwise-normal body. The most commonly reported organisms to experience this rare condition are reptiles, typically snakes. However, the condition has been seen in species across the animal kingdom and even with some human examples. Scientists aren't completely sure why the developing embryo doesn't completely split and instead continues to grow. Possible explanations for cases in reptiles in particular are fluctuations in external temperature during incubation, environmental pollution or exposure to chemical mutagens. Ionising radiation can be an extremely damaging external mutagen. The exposure of

People in Tibet have evolved to live thousands of metres above sea level

Did you know? On average, parents pass 60 new mutations to their children

Types of mutations

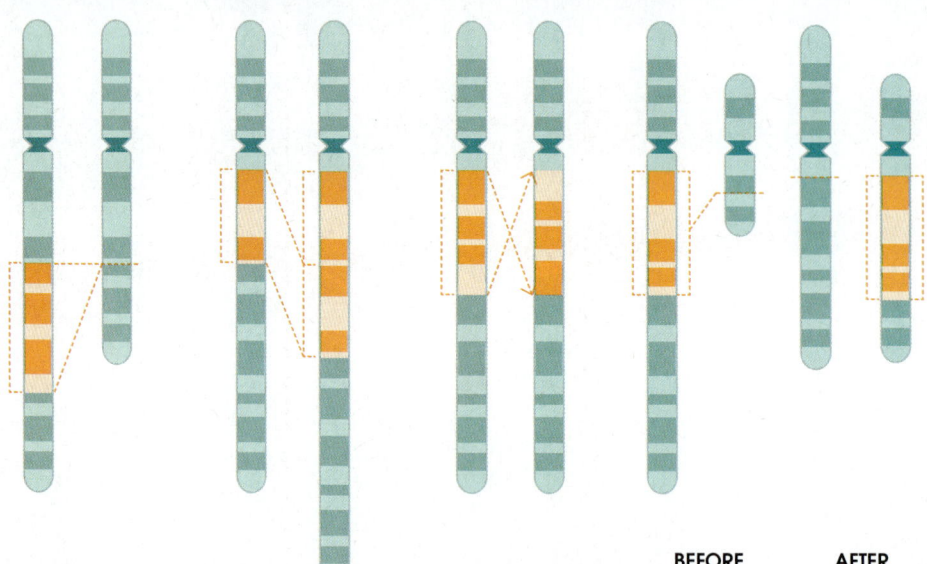

BEFORE AFTER

1 Deletion
This type of mutation occurs when one or more nucleotides from a portion of a DNA strand are lost. Deletion can lead to many genetic diseases, such as cystic fibrosis.

2 Duplication
This form of genetic mutation involves copying a DNA segment and repeating the same segment again before moving onto the next new segment.

3 Insertion
Rather than losing or repeating segments, this involves new nucleotide base pairs being added into a sequence. This type of mutation interprets the creation of vital proteins and leads to different birth defects.

4 Substitution
When one nucleotide is incorrectly swapped out for a different one. This replacement has an effect on the production of a protein, such as disrupting and altering its function or rendering it useless.

THE ANIMAL KINGDOM

MEET THE MUTANTS

living tissue to high or prolonged levels of radiation – such as X-rays and ultraviolet light (UV) – can break down sequences in DNA, damage the structure of cells and lead to their mutation. The result of mutations caused by radiation are typically the development of cancers and growth of tumours, along with some physical abnormalities – just as it was in the wake of the tragic Fukushima Daiichi Nuclear Power Plant accident. In 2011, scientists discovered that the larvae of pale grass blue butterflies (Zizeeria maha), which were exposed to the radioactive fallout of the explosion, developed much smaller wings and irregular eyes compared to unexposed ones. These butterflies showed a mutation rate double that of butterflies found before the incident.

Often referred to as the 'raw materials' for evolution, genetic mutations are the fundamental building blocks for life on Earth.

Mutant fruit

If you've ever bitten into an apple or cut into a tomato and found that the seeds inside have already begun to germinate, you've discovered a mutant fruit. The phenomenon of 'live birth' in some plants is called vivipary. This is where the seeds of plants start to germinate before they have matured and separated from the parent plant. While still nestled within the flesh of a tomato or sandwiched on an ear of corn, a hormone produced by the plant keeps the seeds from entering into the next stage of growth. However, if that hormone runs out, the seeds are free to germinate. In the case of vivipary plants, a genetic mutation inhibits the production of the germination-preventing hormone and tiny shoots prematurely emerge from the fruit.

The seeds of a tomato germinating while still in the fruit, a process known as vivipary

Radioactive mutation
How exposure to radiation can trigger genetic mutations

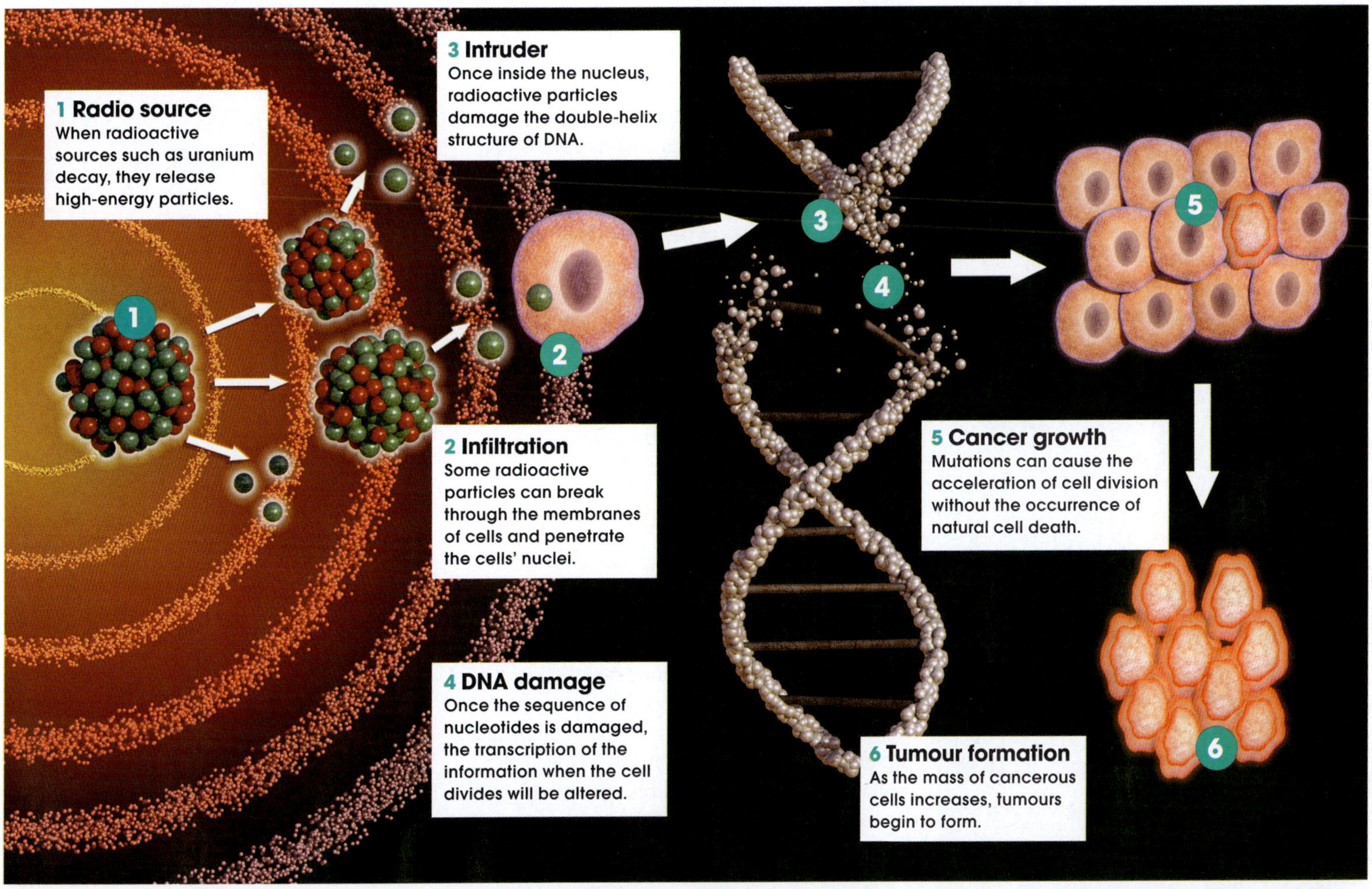

1 Radio source
When radioactive sources such as uranium decay, they release high-energy particles.

2 Infiltration
Some radioactive particles can break through the membranes of cells and penetrate the cells' nuclei.

3 Intruder
Once inside the nucleus, radioactive particles damage the double-helix structure of DNA.

4 DNA damage
Once the sequence of nucleotides is damaged, the transcription of the information when the cell divides will be altered.

5 Cancer growth
Mutations can cause the acceleration of cell division without the occurrence of natural cell death.

6 Tumour formation
As the mass of cancerous cells increases, tumours begin to form.

UNDERSTANDING BIOLOGY

Albinism is a genetic condition where an organism doesn't produce melanin for colouring skin, hair and eyes

The father of evolution Charles Darwin penned the theory of natural selection in the mid-1800s, offering explanations for life and the biological journey it had undertaken through time to become what we know today. The theory of natural selection proposed that only members of a species that are able to adapt to their environments will survive long enough to reproduce and continue their lineage. However, Darwin remained in the dark about the role genetics played in evolution. That knowledge came from Dutch botanist Hugo de Vries at the beginning of the 20th century.

Natural selection is driven by mutations in an organism's DNA that cause a beneficial physical or behavioural trait that ensures their survival. Those traits are then passed on to the next generation, and over time the species evolves. Members of the species that don't genetically mutate and adapt often die before they can pass their DNA on to their offspring.

For an entire species to evolve, the development and mutation of DNA occurs over thousands of years. For example, over millennia the DNA of some giraffes mutated to allow them to grow long necks and tall legs to reach parts of trees that other shorter giraffes and herbivores struggled to get to. Their long necks gave them an advantage and increased their chance of survival, although natural selection and mutation-driven evolution isn't a thing of the past. There are many examples of modern-day mutants, showing scientists the next potential stages in their evolution, including humans.

Two-headed snakes and poison-protected mice are far from being alone in their mutations. To find one of the most advanced mutants on Earth, you need only look in the mirror. Considering the fact that genetic mutations are the raw materials of evolution, we as humans have undergone countless changes in our genetic makeup over the past 10,000 years to become the apex predators and technological innovators that we are today. Over millennia, both small and large mutations to the human genome have propelled our evolution and created the modern human, from the

Did you know?
Ribonucleic acid (RNA) genomes can also mutate

A giraffe's long neck is both a strange and extremely useful adaptation

MRSA
Staphylococcus aureus

As single-celled organisms, each time bacteria create a copy of themselves through cellular division they run the risk of mutation. In 20 minutes or less, bacteria can double their population. When a mutation occurs it quickly becomes the norm, such as in methicillin-resistant Staphylococcus aureus (MRSA): mutations have given it the ability to resist the effects of antibiotics.

THE ANIMAL KINGDOM

MEET THE MUTANTS

MODERN-DAY MUTANTS

Mutations have allowed these species to adapt and survive

European house mouse
Mus musculus

Warfarin is a blood thinner used in pest control. The poison works by reducing vitamin K production, which helps the body form blood clots – an important process that prevents excess bleeding. A mutation in a gene called VKORC1 allows certain mice to produce more vitamin K and combat the blood-thinning ability of warfarin.

Green anole
Anolis carolinensis

In only 20 generations, green anole lizards have rapidly evolved to best a group of invasive brown anoles in the US. Within the span of 15 years, these tree-dwelling lizards have evolved larger toepads with more sticky scales that allow them to reach heights that their invasive competitors are unable to access.

Table coral
Acropora

Warming temperatures cause coral species to 'bleach', where they lose their vital algae and die. Research has shown that table corals have mutated to adapt to warmer waters. Heat-tolerant corals like these are more likely to survive in the face of a warmer climate and may evolve into the dominant type of coral in our oceans.

Peppered moth
Biston betularia

These moths are naturally speckled white and black. During the rise of the Industrial Revolution, they were able to adapt and survive predation thanks to a mutation that emerged around 1819. It made these moths darker – almost completely black – allowing them to blend in on urban tree trunks that had been coated in soot.

UNDERSTANDING BIOLOGY

shedding of our body hair to the development of large brains. One example of a recent human mutation happened between 6,000 and 10,000 years ago, affecting a gene called OCA2 that halted the production of the melanin pigment in the iris that colours the eyes brown, diluting eyes to a shade of blue.

One of the ways humans have evolved – and continue to evolve – relates to our diet. Humans are one of the only mammals that drink milk after infancy. The ability to digest dairy in adulthood is down to a group of genes that allow our bodies to break down the sugar in milk, lactose. In other mammals the genes that codes for proteins that can digest lactose typically switch off after infancy, but a mutation in some humans stops that from happening. "We know there are four independent mutations that we see in various human populations, and they sit in the promotion of this gene," says Laurence Hurst, professor of evolutionary genetics at the University of Bath. "Rather than it being switched off, it gets switched back on again." Over the last 8,000 years, this has led to around 90 per cent of Northwestern Europeans having the ability to digest lactose. It's often thought that humans have jumped off the evolutionary train and have reached their final biological destination, likely due to advancements in medicine reducing the need for natural selection. However, this isn't the case. Adaptive mutations are still prevalent in our species, shaping the way humans evolve.

A royal python with bicephaly, a genetic mutation that causes a single organism to have two heads

Polymelia is a mutation that causes an organism to grow extra limbs, like this four-legged chicken

THE ANIMAL KINGDOM

MEET THE MUTANTS

The ever-evolving human

How some members of our species from around the world are evolving

1 Altitude tolerance
In less than 3,000 years, some Tibetans and Han Chinese people have evolved to live at altitudes most people could not – at altitudes of 4,000 metres where the oxygen levels are around 40 per cent lower than at sea level, Tibetan villagers have been thriving in the thin air. Researchers have identified 30 genetic mutations within those people that live at these impressive altitudes, half of which are related to the way the human body uses oxygen and helps the individual manage haemoglobin concentrations.

2 Deep divers
The deep-sea-diving Bajau people of Southeast Asia have shown a remarkable example of human evolution by building the biological equivalent of a scuba tank. The human spleen is typically used to filter blood as part of the body's immune system. Through genetic mutations, the people of Bajau have evolved a larger spleen than most, around 50 per cent bigger on average. A large spleen can act as a reservoir of oxygenated blood while diving, allowing the Bajau people to deep dive up to 70 metres below the surface.

3 Getting taller
In the mid-18th century, the average Dutch soldier was recorded to be 165 centimetres tall. However, the modern-day Dutchman is a towering 182.5 centimetres tall on average. Over the same period, the average American has grown by only six centimetres. It's thought that the 20-centimetre increase is down to natural selection – a preference for tall men among Dutch women.

4 vegetarian nutrition
Scientists found that 70 per cent of vegetarians in Pune, India, that they surveyed have a genetic mutation in the FADS2 gene that allows them to produce essential omega-3 and other fatty acids from a non-meat diet. Fatty acids such as omega-3 are abundant in fish and red meat, but sparse in vegetables. The FADS2 mutation is thought to help vegetarians effectively process these fatty acids from alternative sources.

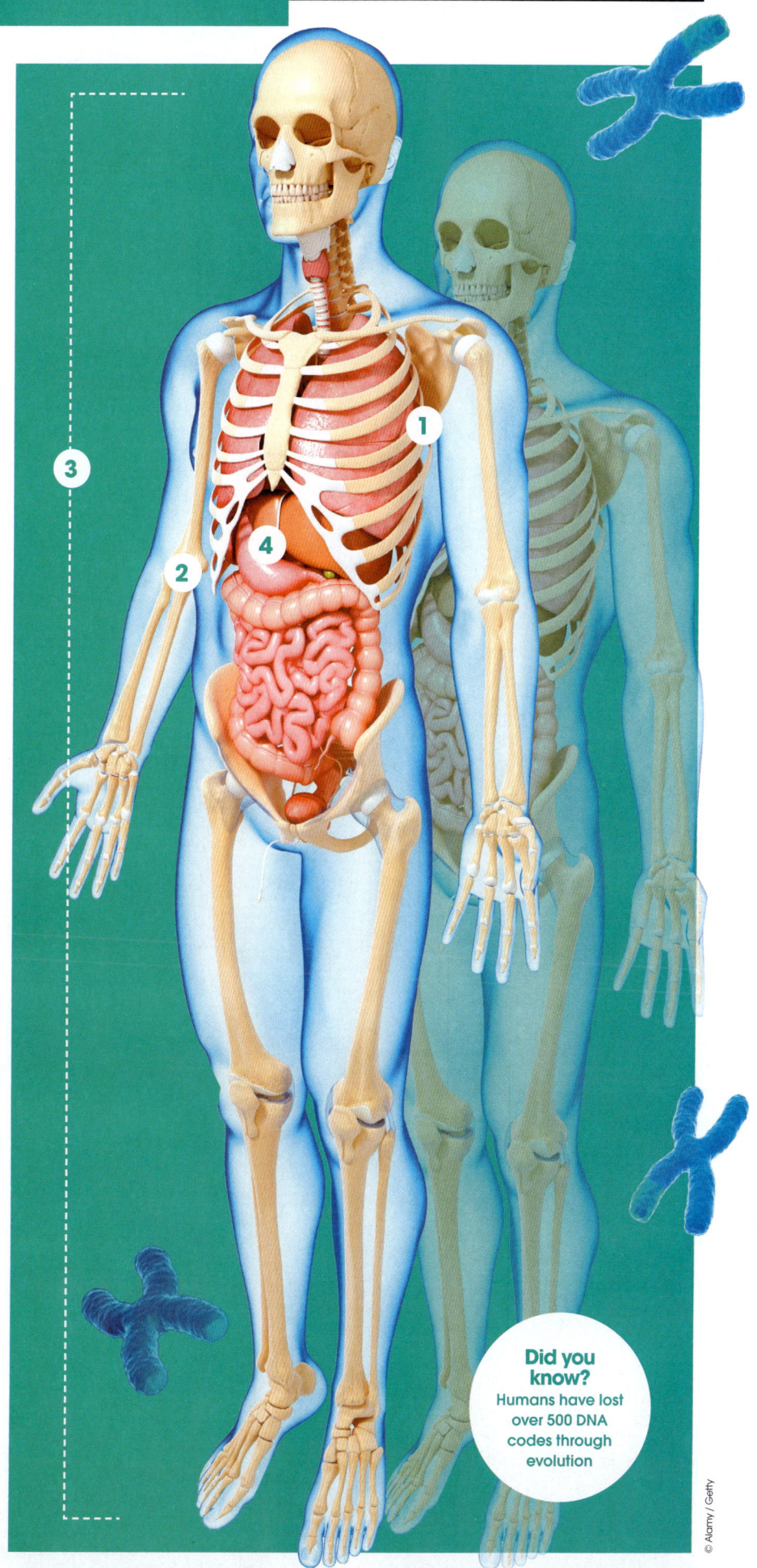

Did you know? Humans have lost over 500 DNA codes through evolution

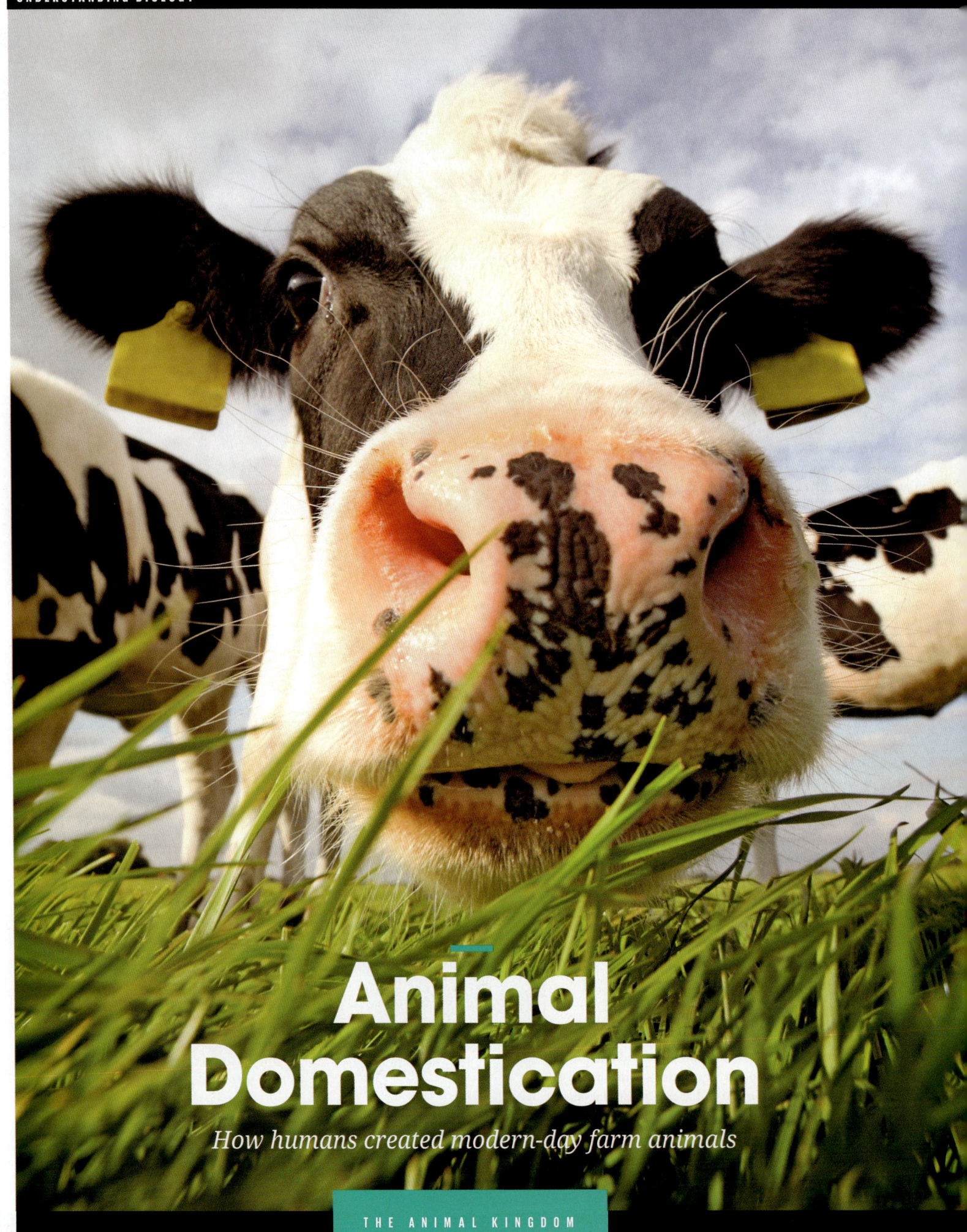

Animal Domestication

How humans created modern-day farm animals

THE ANIMAL KINGDOM

ANIMAL DOMESTICATION

It's fair to say that the domestication of animals is one of the most important advancements in human history. Known as the Neolithic Revolution, around 12,000 years ago humans began cultivating the land and breeding animals for livestock. The principle of livestock domestication centres around selectively breeding wild animals with traits beneficial for humans – for meat, milk or materials. Over millennia, humans have bred generations of animals like cows, sheep, pigs and chickens to enhance qualities such as their muscle mass for meat and wool production for fabrics.

The first wild animal to be taken by humans from its natural environment was the predecessor of the modern-day sheep, the Asiatic mouflon (Ovis orientalis), around 11,000 years ago. Since then, humans have captured and domesticated 38 different species, including 8,800 different breeds, from around the world.

The specific techniques that primitive farmers used to tame and domesticate livestock throughout history are largely unknown to archaeologists. However, in 2014 researchers uncovered some clues in an ancient settlement in Turkey. Around 11,000 years ago, a small village on the banks of the Melendiz River in Turkey was home to some of the earliest domesticated animals. By studying the changes in bones at the site, known as Aşıklı Höyük, archaeologists noticed a shift in the remains of wild animals such as hares, deer and goats, to predominantly sheep remains. By around 9,500 years ago almost 90 per cent of the bones being left at the site were from sheep, 58 per cent of which were from females, typical among flocks for breeding purposes. The remains were found in areas researchers believed to be pens in the middle of the village, used to acclimatise the sheep to humans. Villagers likely introduced young sheep as pets to their villages, too.

The theory as to why wild sheep were originally stabled is related to other agricultural processes, such as crop growing. As prolific hunters, early farmers may have had to weigh up spending time further afield hunting compared to time spent on crop farming. To optimise their time, hunters brought their prey to the farm to breed and produce a sustainable supply of food to the village while still tending their crops.

Animal domestication and human evolution are intertwined. Not only have humans bred livestock and changed the biology of domestic animals, but domestic animals have also affected how humans have evolved. One of the most apparent examples lies in our ability to digest a sugar called lactose. As infants, many mammal species begin their lives suckling milk from their mothers. However, at varying

Did you know?
The average milk yield in the UK is 8,214 litres per cow per year

> **"The domestication of animals is one of the most important advancements in human history"**

Human-made animals

Cats, dogs and cattle aren't the only animals that humans have altered through domestication. Without human intervention, animals such as pigeons and goldfish wouldn't exist. More than 400 million pigeons live around the world, and they all descend from a domestic species called the rock dove (Columba livia). These birds originated in ancient Mesopotamia and Egypt and were enticed into the city as a source of food around 10,000 years ago. Throughout history, rock doves – and subsequently pigeons – have proven to be more beneficial than a quick meal thanks to their navigational skills as messenger pigeons. Naturally, captive pigeons found their way out of breeding centres and into cities, benefiting from the cliff-like buildings for their nests and bountiful sources of food. Now, in an age where pigeon pilots are no longer needed for sending messages, the world is left with a species of bird that wouldn't have existed without humans.

Chickens descended from red junglefowl (Gallus gallus)

Modern-day pigeons are the result of escaped worker pigeons

UNDERSTANDING BIOLOGY

points in time infants are weaned off their milk supply, losing their ability to process lactose and either becoming plant-munching herbivores or flesh-eating predators. Humans, on the other hand, developed a tolerance for the milk sugar, known as lactase persistence. A genetic mutation, lactase persistence enables the continued production of an enzyme that can break down lactose and metabolise it.

Although the ancient ancestors of modern cows, called aurochs, were originally domesticated solely for their meat around 10,000 years ago, around 6,000 years ago dairy farming began among European farmers, roughly around the same time humans developed a tolerance for lactose past infancy. Now lactose persistence is found in around 35 per cent of adults globally.

But domestication isn't all it's cracked up to be. Although raising livestock has facilitated the advancement of human civilisation, it's also brought with it disease. During the evolution of animal domestication, confining animals has facilitated the transmission of pathogens and parasites to humans – for example tuberculosis among cows and influenza from pigs.

Alpacas were first domesticated in the Puna region of Peru

Then and now
What livestock looked like before humans domesticated them

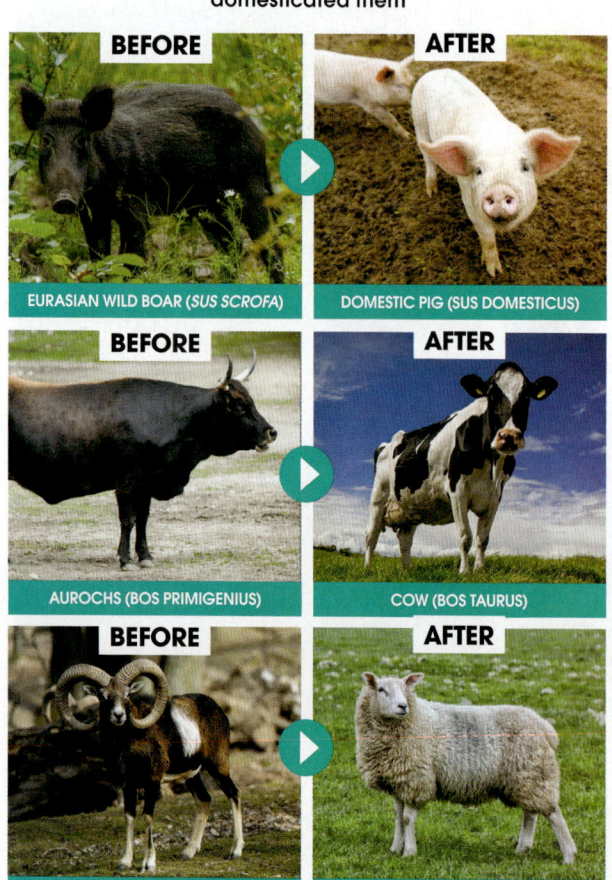

BEFORE — EURASIAN WILD BOAR (*SUS SCROFA*)
AFTER — DOMESTIC PIG (SUS DOMESTICUS)

BEFORE — AUROCHS (BOS PRIMIGENIUS)
AFTER — COW (BOS TAURUS)

BEFORE — ASIATIC MOUFLON (OVIS ORIENTALIS)
AFTER — SHEEP (OVIS ARIES)

Domestication syndrome

Why is it that so many domesticated animals look so far removed from their wild ancestors? The answer lies in a phenomenon called domestication syndrome. When you look at any domesticated animal, you'll notice a lot of them share common physical and behavioural traits, including floppy ears, tameness and changes in their tails. Scientists have been studying this since Charles Darwin proposed it in 1868. Over decades, Russian scientist Dmitry Belyaev conducted an experiment where silver foxes were selectively bred for tameness. After 45 generations of selecting for this behaviour, the foxes had become more like dogs, with wagging tails and floppy ears. The syndrome remains largely a mystery, but some scientists have suggested that selecting for tameness affects cell development during the embryonic stage.

One of the silver foxes from the Russian domestication experiment

THE ANIMAL KINGDOM

ANIMAL DOMESTICATION

Domestication around the world

Where did our farm animals come from?

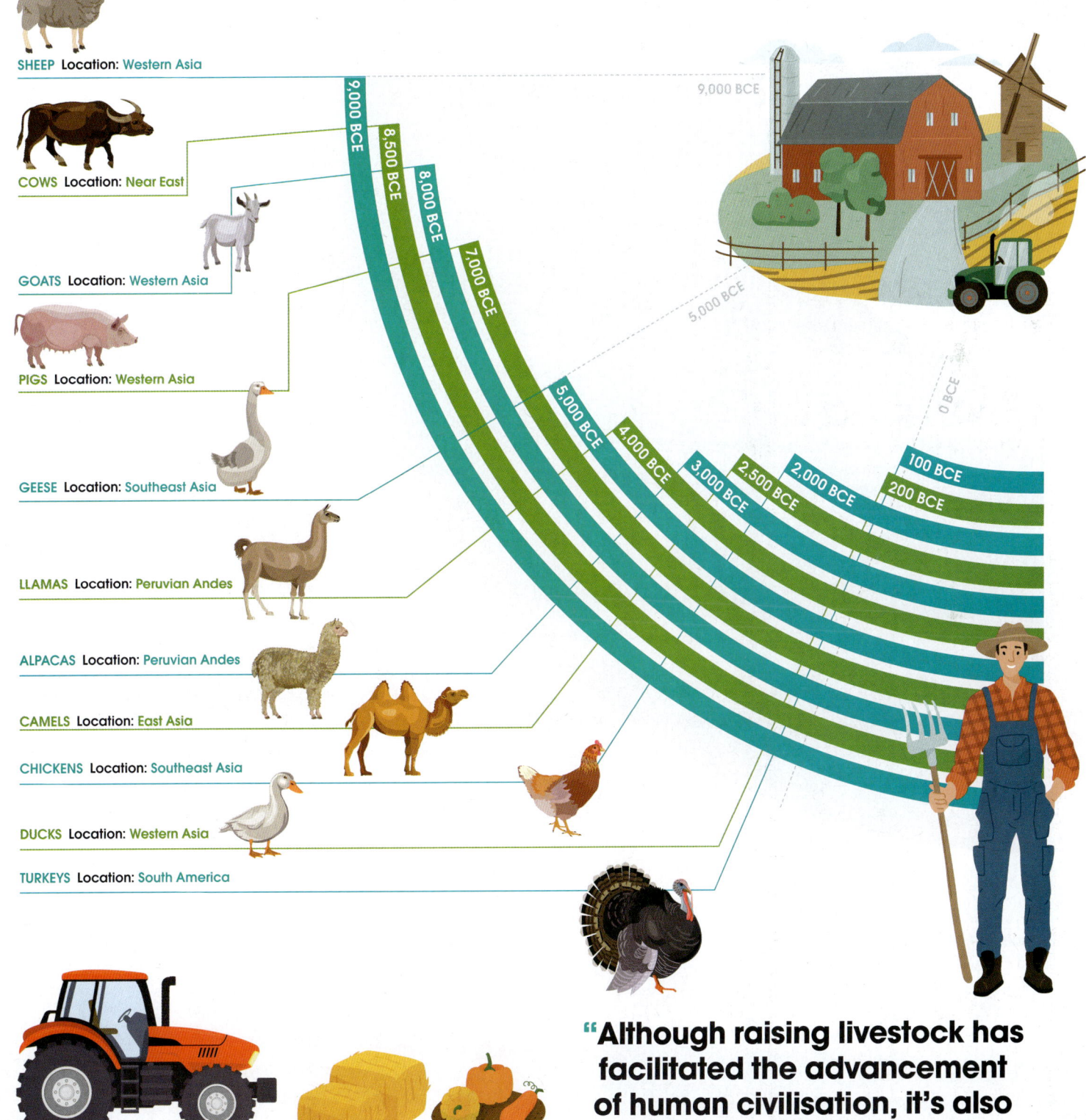

SHEEP Location: Western Asia — 9,000 BCE
COWS Location: Near East — 8,500 BCE
GOATS Location: Western Asia — 8,000 BCE
PIGS Location: Western Asia — 7,000 BCE
GEESE Location: Southeast Asia — 5,000 BCE
LLAMAS Location: Peruvian Andes — 4,000 BCE
ALPACAS Location: Peruvian Andes — 3,000 BCE
CAMELS Location: East Asia — 2,500 BCE
CHICKENS Location: Southeast Asia — 2,000 BCE
DUCKS Location: Western Asia — 200 BCE
TURKEYS Location: South America — 100 BCE

"Although raising livestock has facilitated the advancement of human civilisation, it's also brought with it disease"

UNDERSTANDING BIOLOGY

Why animals shed

What prompts some creatures to regularly lose their old outer layers and replace them with fresh new skin?

Many creatures throughout the animal kingdom undergo a process of physical transformation called moulting. Whether they create ghostly serpent sleeves as a result or just ditch clumps of old skin and fur, shedding is an important stage in an animal's growth and survival. The terms moulting and shedding are often used interchangeably, but there's a slight difference between each process. Moulting is the removal of an animal's entire skin, feathers, shell or exoskeleton as it grows, which often occurs on a seasonal basis. For example, elephant seals (Mirounga leonina) undergo what is known as a 'catastrophic moult' during late summer. As dramatic as it sounds, the 'catastrophic' nature of these moults merely relates to the large patches of skin that shed at one time. During this time, blood flow within the seals is redirected towards their skin to produce a new outermost skin layer, known as the epidermis. This puts their vital organs at risk in the freezing oceans they inhabit, so moulting has to be carried out on land.

Shedding hair and skin, on the other hand, can occur more regularly and be the result of dryness or a temperature change. Snakes generally moult

THE ANIMAL KINGDOM

WHY ANIMALS SHED

After around 30 days, king penguin (Aptenodytes patagonicus) chicks moult their juvenile feathers

Did you know?
Canadian geese have around 25,000 feathers on their body

A human's best hypoallergenic friend

Along with a heap of excess fur to clean up, being around a dog that likes to shed its fur can trigger human allergies. Siberian Huskies, for example, are some of the heaviest shredders in the canine world. This is mainly due to their role as sled dogs in cold climates, as shedding helps maintain their thick, healthy coats. But what's good for plummeting temperatures can wreak havoc on their owner's nose. When dogs shed, they also shed skin cells. These can find their way into the nostrils and mouth of the owner, triggering the body's immune system and causing allergic relations. However, many dog breeds, such as the Afghan Hound and Giant Schnauzer, have been deemed hypoallergenic. This means that while no dog is completely free from shedding, some shed at such a low level that allergic reactions can be avoided.

The coats of Giant Schnauzers don't often shed, making them hypoallergenic

UNDERSTANDING BIOLOGY

dog breeds shed their fur all year round. Even humans unknowingly spend their time shedding old skin and hair – around 500 million skin cells and around 100 hairs each day.

Snakes are some of nature's best moulters. In the same way that humans outgrow clothes through their early years of life, snakes periodically peel a surface layer of scales to prepare for their bodies to expand and grow. Between 4 and 12 times a year, a snake will undergo a full-body moult, known as ecdysis. The outer layers of skin cells are detached in a continuous sheet from the newly grown dermis below. A protective scale covering the snake's eye, called the spectacle, also sheds during this process. While all snakes and reptiles shed, not all of them create fully formed scale replicas of themselves. Tortoises, turtles and some lizards shed their skin gradually, often in dry patches that flake away.

Like snakes, there's a whole host of animals that undergo complete ecdysis when it's time to grow up, including eight-legged arachnids, crustaceans and all manner of insects. Unlike the cocoon-spinning abilities of many metamorphic insects, such as moths and butterflies, others make their winged transformations beneath their protective exoskeletons. As larvae, dragonflies live beneath the water, often found in ponds and lakes. These wingless juveniles skip the pupal stage of metamorphosis and head straight to moulting.

> **Did you know?** Some bears exfoliate with rocks during moulting

After several days of internal metamorphosis underwater, the larva climbs out of the depths and up onto a nearby leaf, and within a few hours the newly winged dragonfly breaks through the beak of the previous exoskeleton, called an exuvia. But what happens to the leftover exoskeleton?

Some insects, like dragonflies and cicadas, don't do anything with their exuviae. However, many insects, such as pill bugs (Armadillidium vulgare) and American cockroaches (Periplaneta americana), waste no time in munching down on their previous bodies to reabsorb valuable proteins and chitin within the exuviae. Leaving behind a hollow replica of your body can also act as a beacon to predators, signalling that there's prey around that may be without its protective exoskeleton. Eating their old body reduces the risk of being discovered.

Reptiles like chameleons shed their skin in patches

Eating your own skin

For amphibians, skin is much more than a protective barrier – it's also how they breathe. Frogs and toads have both evolved moist skin that allows gases such as oxygen to pass through the surface and into their bloodstream. As amphibians grow, their skin sheds, but not as a whole piece like their reptilian cousins. Instead, patches flake away from the body regularly, with some species shedding their entire skin daily and others taking a weeks to completely shed. Once the old skin layer has separated, amphibians ferry the flake along the body towards the mouth and eat it. Known as dermatophagy, the process of regularly eating their skin not only lets the amphibian recuperate some of its nutritional losses but also prevents it from drying out. Ensuring that their skin is soft and permeable enough to exchange atmospheric gases is vital to their continuing survival.

> **"A snake will undergo a full-body moult, known as ecdysis"**

At the start of their moult, snakes shed a scaly cap from their eyes

A green tree frog (Hyla cinerea) feeding on its own moulted skin

WHY ANIMALS SHED

Scale separation

How snakes peel away their skin and leave a scaly sleeve behind

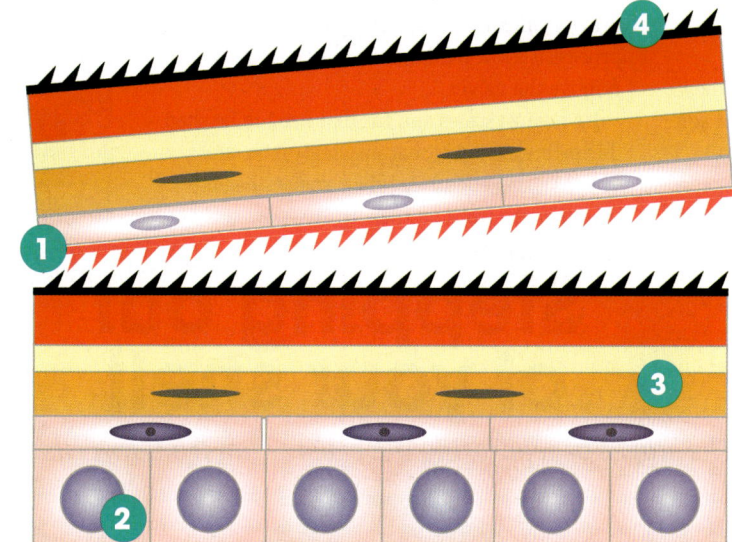

1 Hormones
Days before shedding begins, hormones are released throughout the snake's body that trigger ecdysis.

2 Cell proliferation
During moulting, all the cells in the snake's outermost layer, called the epidermis, rapidly increase in numbers.

4 Terminal differentiation
The old epidermis breaks away from the hardest surface layer of the snake's skin, called the oberhautchen.

3 Differentiation
Cells called keratinocytes change to form new layers that make up the snake's epidermis.

5 Eyes
During the moulting stage, the eyes first turn a milky blue colour as the spectacle cap separates.

6 Head stretching
With its vision impaired, the snake carefully rubs its head on an abrasive surface to form a tear in the epidermis.

7 Snake skin undress
Crawling through tight spaces and weaving through rocks and plants causes the epidermis to peel away from the body along the tear.

8 Old skin
Once the snake has wriggled its way out of its old skin, what remains is an inside-out serpent sleeve made from both robust keratin scales and stretchy epidermis cells.

Like many animals, birch trees shed their bark to remove parasitic pests

29

UNDERSTANDING BIOLOGY

Instead of shedding an entire body's worth of skin and exoskeleton, some animals, such as reindeer and moose, undergo a partial moulting process to reveal a new set of antlers. Unlike the horns on the heads of buffalo or gazelle, reindeer have similar structures called antlers. Antlers are bony appendages that grow out of the skull, as opposed to horns, which are not attached to the skull and are predominantly made of the same material as human hair and nails, called keratin. As antlers grow, they emerge from the head covered in skin and soft hair, commonly called velvet. As the animal grows, the bone within develops and compacts to form rigid antler bone. When the antlers are fully developed, the velvet dries out and begins to feel itchy. This causes the antler owner to scratch them against hard surfaces such as trees and rocks to tear away the skin and velvet, leaving only the bone behind. This stage of a reindeer's development might look like a scene from a gory movie, but blood supply is reduced during shedding, and it's believed that this prevents the animals from feeling any pain. The process is swift – it takes around 24 hours to remove the velvet.

It's not just scales and fur that need to be shed during the changing seasons. Before the chill of winter descends, many bird species begin maintenance work on their bodies, particularly

The exuvia of a cicada is a perfect copy of its former body

Stepping out of your shell

The stages of separating the crustacean from its crust

1 Rupture
Ecdysis begins when a membrane connecting the crab's thorax and abdomen ruptures.

2 Head separates
The head of the crab divides from the thorax and the abdomen, revealing the start of a new head.

3 Abdomen splits
The abdomen underneath the crab breaks away from the new body, creating enough space for the crab to escape the old shell.

4 Slipping out of the legs
The legs of the crab, called the pereiopods, are pulled free from the old shell.

5 Final push
In the last phase of ecdysis, the abdomen is fully withdrawn.

6 Free at last
Once the claws, also known as the chela, are free from the shell, ecdysis is complete.

THE ANIMAL KINGDOM

WHY ANIMALS SHED

Deer strip away the itchy velvet from their antlers

Hermit house-swapping

There are more than 800 different species of hermit crabs throughout the world's oceans. Like other crustaceans, hermit crabs undergo ecdysis and moult their bodily shells. Hermit crabs also need to replace the gastropod shells that they call home when they've outgrown them. Once a larger, unoccupied shell has been spotted, the small crustacean releases its old shell and backs into the new one. Hermit crabs use the tip of their adapted abdomen to clasp onto a pillar-like structure found within many shells, called the columella, to secure themselves in place. Although crabs often make the switch when they encounter a rogue shell, crabs such as long-wristed hermit crabs (Pagurus longicarpus) exchange houses in a chain with other crabs. When these exchanges aren't amicable, crabs will attack each other by striking each other's shells in the hope of securing a shell eviction.

If a hermit crab can't find a shell that fits, they'll reduce their size between moults

their feathers. Throughout the year, some feathers may have suffered wear and tear, and so the birds embark on patterns of shedding, though this can come at a cost to their ability to fly. During moulting, some birds, such as swans and geese, lose all of their flight feathers at the same time, completely grounding them for around six weeks before they can grow new ones.

Other than outgrowing their own body, there's one other reason an animal might want to step outside of their old skin... parasites. The natural world is full of fungal, bacterial and animal parasites whose main goal is to find a way to extract sustenance from their host. The ocean is littered with parasitic organisms that are just waiting to find their way into the cracks of a crustacean's shell. Parasites such as isopods and barnacles try to burrow their way past the protection of crab and lobster shells to syphon off nutrients and find a source of food. To prevent parasites from taking hold, crustaceans undergo ecdysis to grow a new body within their existing shell and then step out of their old one, hopefully leaving the parasites behind.

Creating a new shell or exoskeleton comes at a cost to the creature. Once a giant mud crab (Scylla serrata) has completed ecdysis, the body that emerges from the previous shell is completely soft. While the crab is vulnerable to predation without its usually tough exterior, it can find a protected spot and wait for its new shell to harden. Within a matter of days, the new shell hardens and the crab is free to face the world again.

It's not just critters and crabs that shed to strip away parasites, with some plant life doing the same. Throughout late spring and early autumn, the silver birch tree (Betula pendula) sheds thin paper-like layers of its bark. This reduces the opportunities for parasites and other invertebrates to burrow into healthy bark and potentially cause an infection. The fallen bark also releases nutrients back into the soil when it decomposes, which can then be reabsorbed through the tree's roots.

But moulting isn't for everyone. Some animals have avoided shedding entirely, despite their growth. Fish scales, for example, grow with the fish rather than being shed and replaced. This can be handy for scientists trying to accurately age them; some fish species have scales with concentric ridges – similar to the rings in the trunk of a tree – that are used as an indication of age.

> **Did you know?**
> Pea crabs have shells just two centimetres long

31

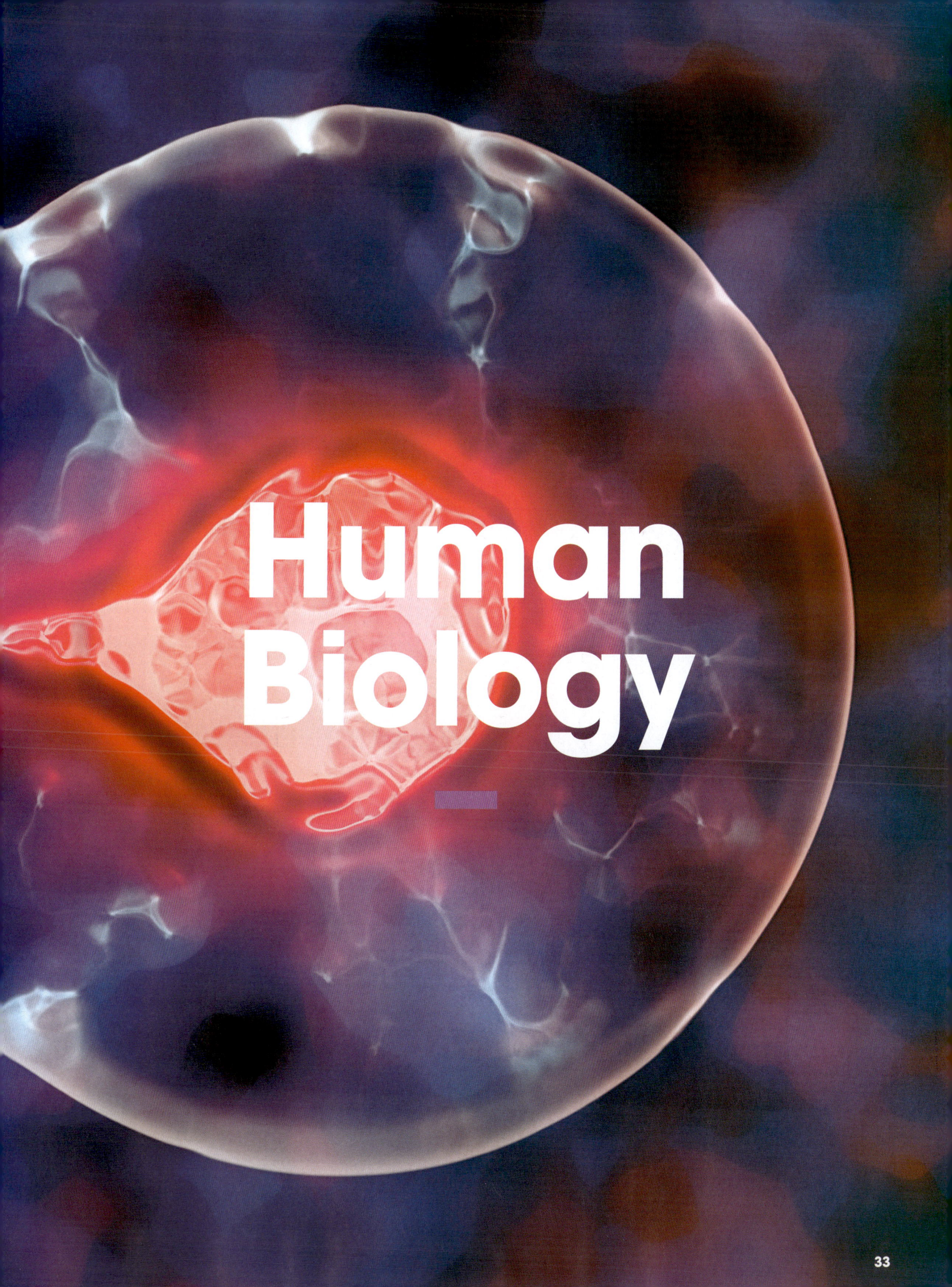

UNDERSTANDING BIOLOGY

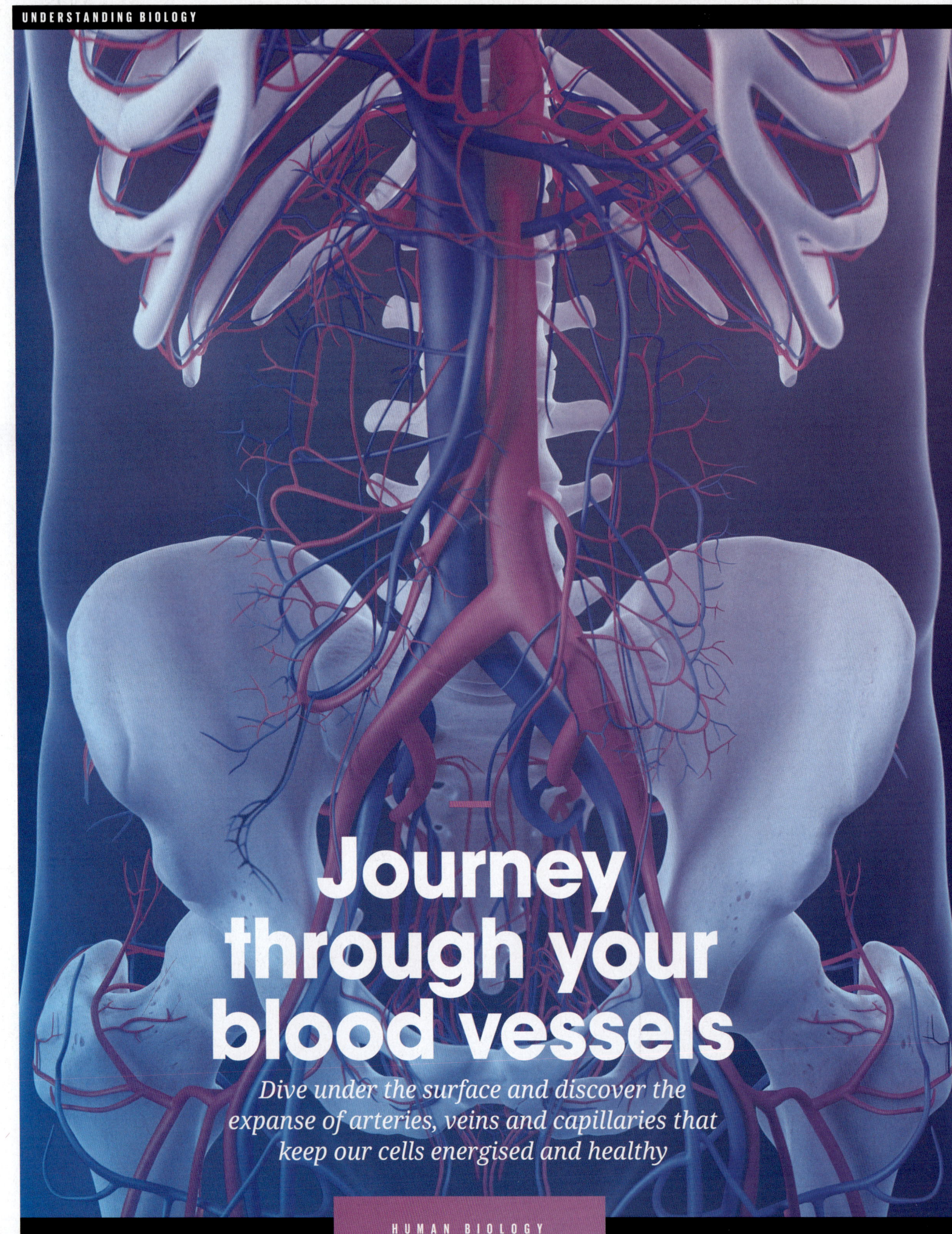

Journey through your blood vessels

Dive under the surface and discover the expanse of arteries, veins and capillaries that keep our cells energised and healthy

HUMAN BIOLOGY

JOURNEY THROUGH YOUR BLOOD VESSELS

Every cell in the human body requires oxygen to produce energy, but most of our cells cannot access it directly. A single-celled life form interacts directly with its environment, exchanging nutrients and waste products as required. But our ancestors gave up that lifestyle hundreds of millions of years ago when they evolved to become multicellular organisms.

As multicellular life grew more sophisticated, our ancestors' cells became specialised and compartmentalised. Then, many millions of years later, their descendants migrated from the ocean to the land. Gone were the days when even external cells exchanged nutrients with their environment – now our ancestors' cells were encased inside a protective barrier of skin, allowing them to retain their water and maintain consistent internal temperatures. This meant that precious few cells interacted with the environment, and therefore very few cells could access much-needed oxygen and sugars for energy. Fortunately, our species – just like our land-treading ancestors – possesses an interwoven network of tissues and organs dedicated to ensuring our cells acquire the nutrients they need. We call this bodily network the circulatory system.

This vascular network consists of a pump – the heart – and a connected network of blood vessels that carry blood to and from internal tissues. If blood and its component parts are the delivery service, busily dispensing oxygen and collecting waste, then blood vessels are the highways and smaller roads on which they travel. Together with the heart, which provides the pressure that propels blood around the circuit in the body, blood vessels are essential for maintaining the health and functionality of our cells. Our circulatory system is also highly adaptable. When we're at rest and require less energy, heart rate slows. However, when we move about and exercise the heart rate rises.

You can feel the beating heart in action for yourself by placing a finger on the left side of your wrist or by softly placing a finger next to the left side of your windpipe. When you do this, you're feeling your pulse through the radial and carotid arteries respectively. As well as the heart, blood vessels themselves react to environmental changes. When it's cold, for example, blood vessels constrict, helping to reduce heat loss. If you've ever suffered brain freeze, blame your protective blood vessels, which constrict as the cold substance hits the roof of your mouth. The reverse is also true, as blood vessels expand when it's hot to help our bodies shed excess heat. As well as supplying the body with gases and nutrients, our vascular network helps regulate our internal environment, helping to both fuel and protect our cells.

Did you know? Blood travels around the body in less than 60 seconds

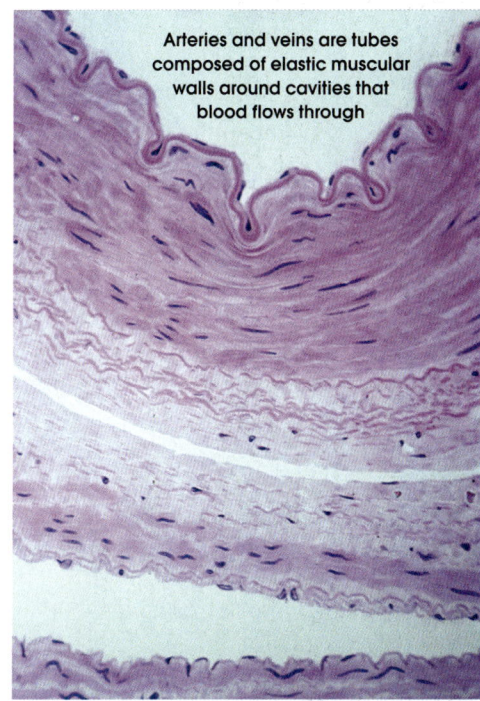

Arteries and veins are tubes composed of elastic muscular walls around cavities that blood flows through

> "An interwoven network of tissues and organs dedicated to ensuring our cells acquire nutrients"

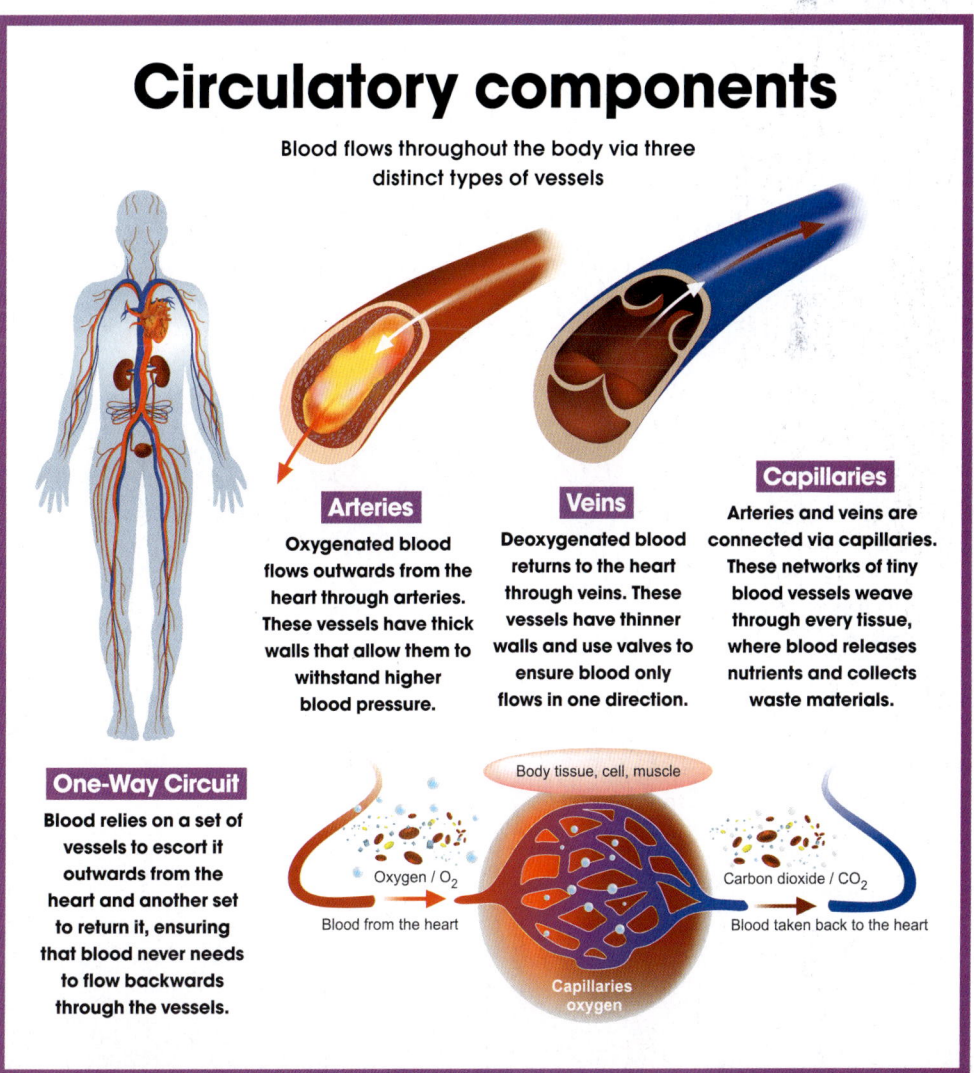

Circulatory components
Blood flows throughout the body via three distinct types of vessels

Arteries
Oxygenated blood flows outwards from the heart through arteries. These vessels have thick walls that allow them to withstand higher blood pressure.

Veins
Deoxygenated blood returns to the heart through veins. These vessels have thinner walls and use valves to ensure blood only flows in one direction.

Capillaries
Arteries and veins are connected via capillaries. These networks of tiny blood vessels weave through every tissue, where blood releases nutrients and collects waste materials.

One-Way Circuit
Blood relies on a set of vessels to escort it outwards from the heart and another set to return it, ensuring that blood never needs to flow backwards through the vessels.

Body tissue, cell, muscle
Oxygen / O_2 — Blood from the heart
Carbon dioxide / CO_2 — Blood taken back to the heart
Capillaries oxygen

UNDERSTANDING BIOLOGY

Circulation during gestation

Humans are placental mammals, which means our offspring acquire their nutrients from a placenta during foetal development. Budding embryos swiftly develop a blood supply within the first few weeks of development, but without functioning lungs, kidneys or a gastrointestinal tract, a foetus must rely on its parent for oxygen and nutrients. The maternal blood supply is connected to foetal circulation via the placenta and the umbilical cord, which contains two umbilical arteries and one umbilical vein. The two arteries escort deoxygenated blood from the foetus to the placenta, while the vein carries oxygenated and nutritious blood from the placenta to the foetal heart.

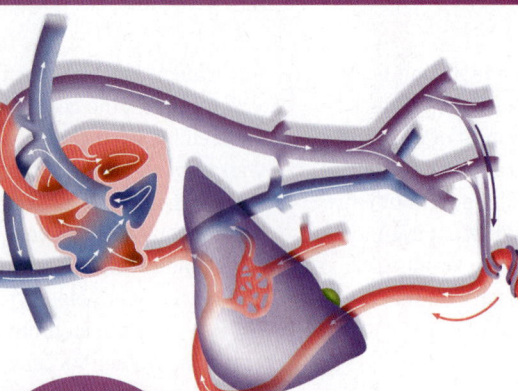

A growing foetus relies on blood exchange with its parent for oxygen and nutrients

Fuelling the body

Follow the flow of blood as it travels through the circulatory system

Did you know? The average adult has about 5.5 litres of blood in their body

1 Setting off
Oxygenated blood leaves the left ventricle via the ascending aorta. The coronary arteries that provide blood to the heart originate at this section.

2 Aortic arch
The ascending aorta feeds into the aortic arch – the next section of the main artery taking blood away from the heart. The brain-fuelling carotid arteries branch upwards from this region.

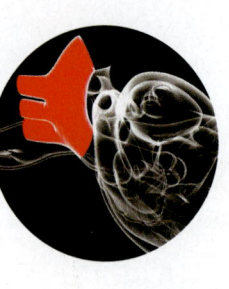

3 Descent
The descending aorta supplies blood to organs in the torso, including the kidneys and gastrointestinal tract, as well as the legs.

8 Resupply
Deoxygenated blood arrives at the right side of the heart and is pumped out towards the lungs through the pulmonary artery.

10 Back to the start
Freshly oxygenated blood returns to the left side of the heart through the pulmonary vein, where the circuit restarts.

9 The exchange
Thin sacks in the lungs called alveoli are surrounded by capillaries, allowing blood to exchange carbon dioxide in the bloodstream with oxygen in the airways.

HUMAN BIOLOGY

JOURNEY THROUGH YOUR BLOOD VESSELS

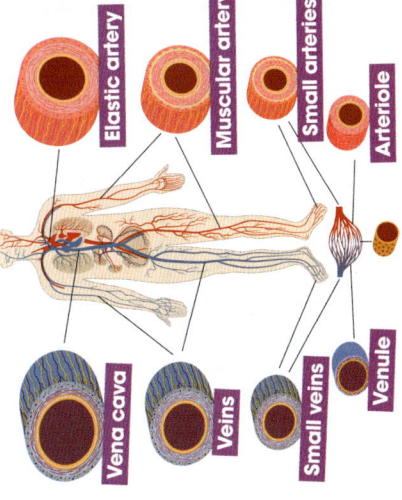

Types of blood vessels

For blood to efficiently migrate throughout and around the body, it must maintain optimal levels of pressure. Oxygenated blood from the left side of the heart is pumped out at high pressure, so arteries must be able to withstand and maintain this force. They withstand pressure by possessing a thick, muscular wall with an outer, middle and inner layer, and they maintain pressure by possessing a narrow lumen – the space that the blood travels through. Elastic arteries, which are found near the heart muscle, possess more elastic tissue in their middle layer. This helps convert the incremental pulses of pressure from heartbeats into a more constant pressure. Capillaries are also highly pressurised, but on a much smaller scale. Their lumens are very narrow and their cell walls are only one cell thick. In contrast, veins also possess three layers in their walls, but these are much thinner. But their lumens are much wider, yielding lower pressures.

4 Into the depths
The common femoral artery feeds the deep femoral artery that supplies blood to the buttocks, femur and hips, and the superficial femoral artery that supplies the lower leg.

5 Branching paths
Blood travels through smaller arteries, into yet smaller arterioles, then into capillaries – the smallest set of blood vessels.

6 Nutrients for waste
Capillaries are thin enough that oxygen and nutrients can be exchanged for carbon dioxide and waste products between the blood and neighbouring cells. Deoxygenated blood then exits the capillaries into venules.

7 The return journey
Smaller venules feed into the major femoral vein, from which blood travels back towards the heart via the inferior vena cava.

> "Our vascular network helps regulate our internal environment"

The abdominal aorta can undergo an aneurysm that causes it to bulge, which can lead to a rupture

© Alam

37

UNDERSTANDING BIOLOGY

Common diseases of blood vessels

In the UK, diseases affecting blood vessels are among the biggest killers each year. The cells of our body are so dependent on the oxygen and nutrients supplied by blood that a blockage or rupture in the vascular network can quickly cause catastrophic damage. While some diseases are genetic, many common diseases are caused at least in part by lifestyle choices, such as a poor diet, which results in the bloodstream carrying more harmful compounds than it ideally should. With an increasingly high-fat and sedentary lifestyle, these diseases are growing ever more frequent. Fortunately, however, lifestyle changes and medical innovations are helping to save lives.

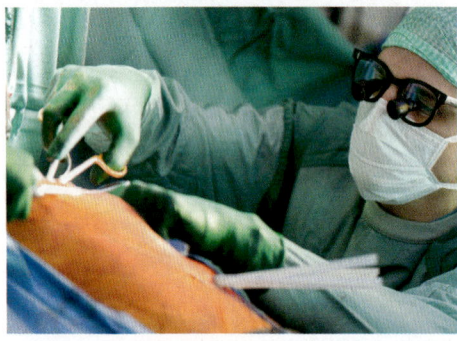

Surgery is sometimes necessary to combat blockages and ruptures of the vascular network

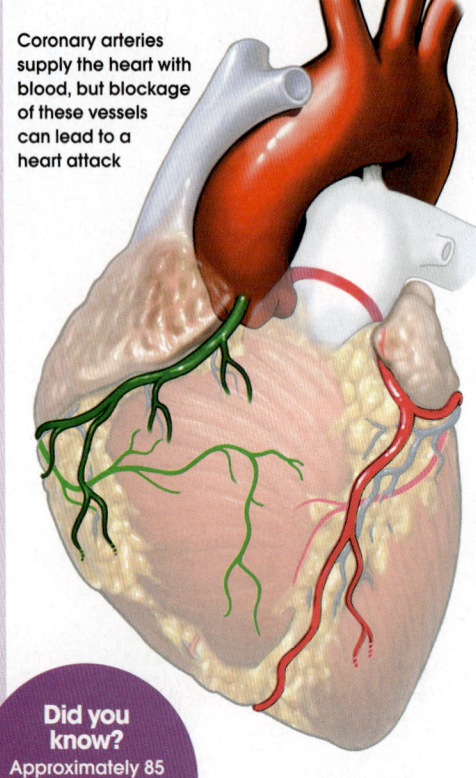

Coronary arteries supply the heart with blood, but blockage of these vessels can lead to a heart attack

Coronary artery disease

Just like all other muscles in our body, the heart requires a supply of blood to fuel itself. However, rather than gaining nutrients and oxygen from oxygenated blood pumped in and out of its internal chambers, the heart muscle relies on coronary arteries wrapped around its exterior for its blood supply. Over time, components transported by the blood such as cholesterol can stick to the walls of coronary arteries, initiating blockages that can partially or completely block the blood supply to parts of the heart. This process is known as atherosclerosis and can result in angina, which is chest pain caused by an insufficient blood supply to the heart. If the coronary arteries are fully blocked, however, cell death of part of the heart can occur, causing a heart attack. Coronary artery disease can be treated using bypass graft surgery, where arteries are rerouted to supply the regions of the heart cut off by coronary artery blockages.

Did you know? Approximately 85 per cent of strokes involve blockages

Fixing vessels

An angioplasty uses inflated tubes to salvage obstructed arteries

BEFORE
Cholesterol floating in the bloodstream becomes attached to the wall of the artery, drawing fats and causing inflammation. This results in the formation of a plaque which narrows the artery.

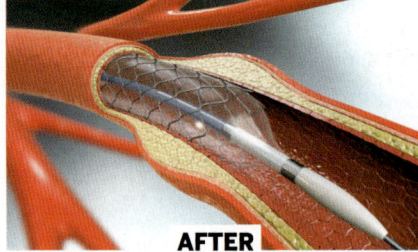

AFTER
A thin catheter is threaded through the artery towards the plaque buildup. A balloon at the tip is then inflated, pressing a mesh wire against the plaque, forcing it outwards and widening the artery.

Stroke

Atherosclerosis describes the accumulation of plaque on arterial walls. These may originate at various places throughout the body, but can become dislodged and carried elsewhere in the bloodstream. Eventually they can become stuck and cause a blood clot, blocking the artery and preventing blood flow to tissues and organs. If this blood clot occurs in the arteries that feed the brain, it can cause an ischaemic stroke. After being deprived of oxygen, brain cells very swiftly begin to die, causing numerous symptoms relating to the areas controlled by the affected region of the brain. If one side of the brain is damaged, the opposite side of the body shows symptoms, which can include drooping of one side of the face, numbness in the corresponding arm and slurred speech.

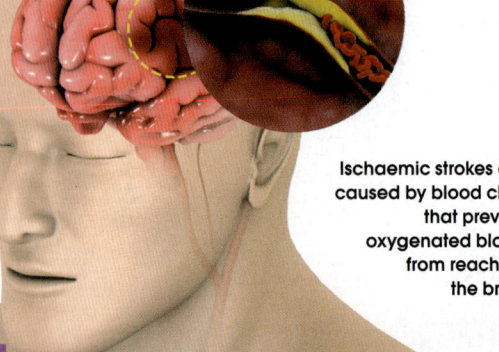

Ischaemic strokes are caused by blood clots that prevent oxygenated blood from reaching the brain

HUMAN BIOLOGY

JOURNEY THROUGH YOUR BLOOD VESSELS

Peripheral artery disease

A narrowing or hardening of artery walls can prove fatal when affecting organs such as the heart or brain, but obstructed arteries feeding the peripheral organs can go unnoticed until the disease is at an advanced stage. Depleted blood flow to the legs can cause symptoms such as cramping, an absence of hair and gangrene that occurs following cell death. Individuals living with diabetes are at particular risk of developing peripheral artery disease due to the abundance of glucose in blood plasma. This interferes with other components in the blood, leading to plaque formation. However, a healthy lifestyle involving regular exercise, not smoking and a low-fat, high-fibre diet can help prevent arterial disease for most adults.

When blood flow is cut off from the feet, cells can begin to die, leading to gangrene

Deep vein thrombosis

Blood clots can appear rapidly in veins where blood flow is slow and laboured, most common in veins deep in leg muscle. During these cases, such as when a person's legs remain unmoved for long periods, an excess of natural clotting factors in the bloodstream can overstimulate the production of a blood clot. In healthy blood, clots form to prevent bleeding, but during deep vein thrombosis the formed clot can grow to a substantial size – large enough to obstruct blood flow. This can cause pain, swelling and a change in colour of the leg. However, the clot can cause more damage if it travels further through the vascular network into the vessels supplying the lungs, resulting in a pulmonary embolism that can prove fatal.

Deep vein thrombosis can lead to a pulmonary embolism, which obstructs blood flow in the lungs

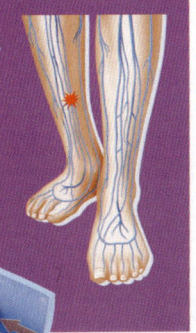

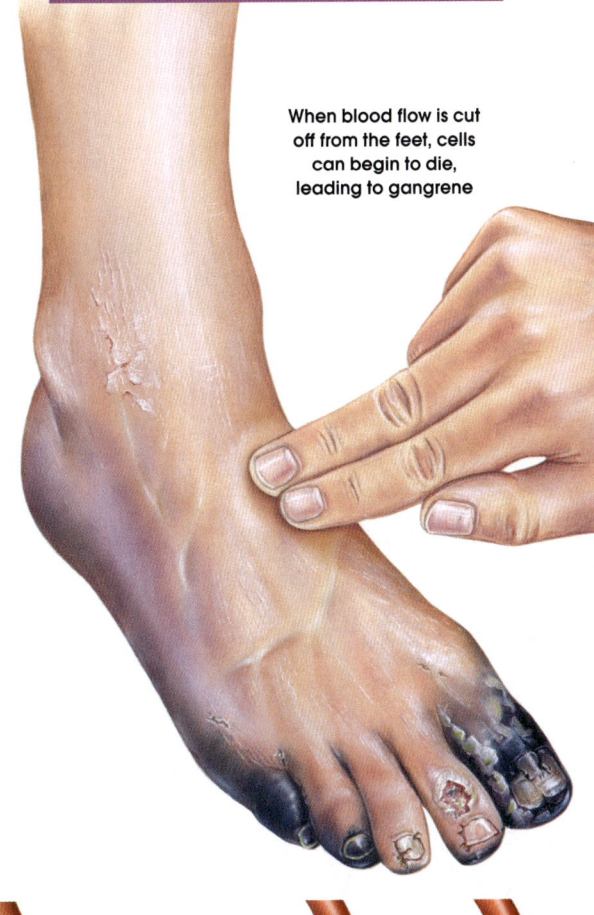

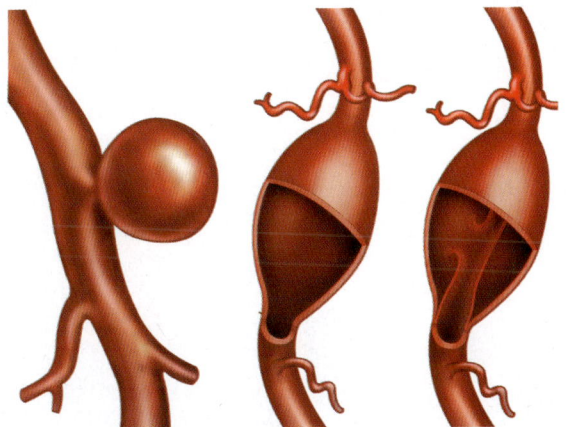

Blood vessel aneurysms can appear in different forms: secular, with bulges on one side (left); fusiform, with bulges on all sides (centre) or dissect, where blood flows into an internal tear, causing a bulge (right)

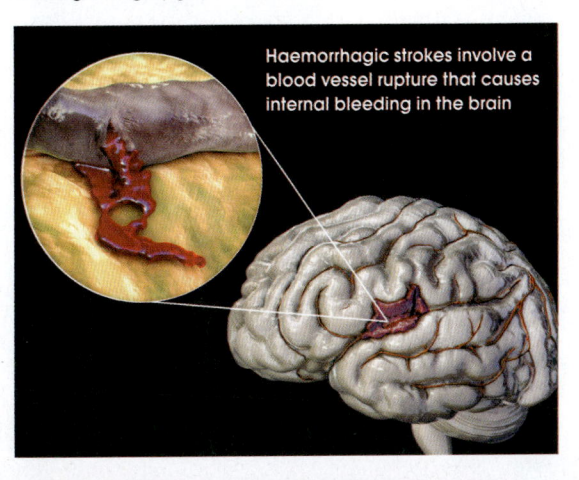

Haemorrhagic strokes involve a blood vessel rupture that causes internal bleeding in the brain

8%
Nearly a tenth of your body weight is blood

62,140 MILES
Total length of blood vessels in the human body

100,000
Estimated deaths from pulmonary embolisms in the US annually

Peritubular cells in the kidneys can detect oxygen deficiencies in the blood

120/80MMHG
The upper end of healthy blood pressure in an average adult

60 TO 100
Average range of heartbeats per minute in an adult

0.005 MILLIMETRES
The width of your smallest blood vessels are one-tenth that of a hair

The liver receives the largest blood supply when the body is at rest

8,000 LITRES
The amount of blood pumped through the vascular network each day

54%
Over half of your blood is found in systemic veins

UNDERSTANDING BIOLOGY

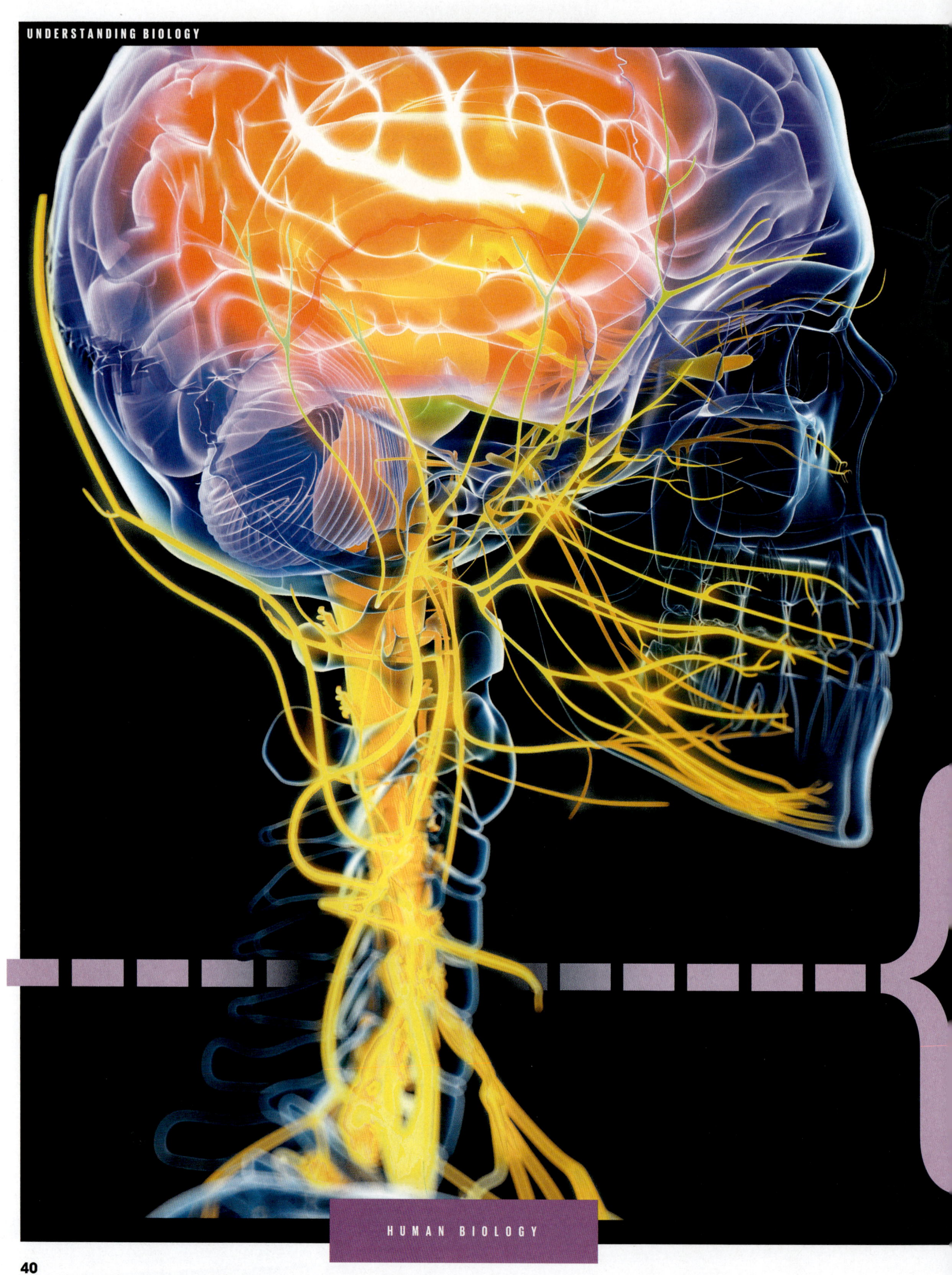

HUMAN BIOLOGY

UNRAVELLING THE
NERVOUS SYSTEM

Unravelling the nervous system

Explore the electrically charged network composed of billions of cells that coordinates your thoughts, feelings and actions from head to toe

Human beings are wonderfully complex. Relative to the numerous single-celled organisms on our planet, humans are gargantuan superstructures. We boast trillions of cells that work to assemble and maintain an array of specialised tissues, organs and bones. Together these form a single being that walks, talks, thinks and feels. This exceptional complexity is only made possible by a coordination centre that monitors and controls the actions of our human form. We refer to this coordination centre as the nervous system. Consisting of the brain, spinal cord and reams of nerves that connect them to the rest of the body, the nervous system is a truly vast and dense network of cells. Collectively, their function is to exchange and relay information through electrical impulses, giving us the power of thought and action.

Many prominent thinkers of Hellenic antiquity, the ancient civilisations around Greece, were obsessed with the complexities of human thought, constantly questioning and probing our understanding of ourselves and our place in the natural world. But some scholars, such as the Greek anatomist Galen, took a literal interest in our thoughts as well as a philosophical one, and decided to delve into dissected pigs in search of the organ that made thought possible. Through his studies, Galen collected evidence that centred the brain as the hub of thought, action and memory. As well as championing the brain as the nexus of the body, Galen also found nerve fibres in his pig subjects that spread outwards from the brain and into the periphery of their bodies. He even went as far to decipher that these nerves were not all the same, but specialised into sensory nerves that received signals and motor nerves that relayed instructions.

Many years later, but not too far away, Renaissance anatomists in Italy sliced and diced human brains to expand our knowledge of the organ. They discovered specialised regions like the pons, which acts as a 'bridge' connecting other brain regions, and the hypothalamus, which was found later to control hormone production. In England in the following century, Thomas Willis deduced that the brain controlled both voluntary and involuntary actions. Scholars were beginning to map the network of the nervous system, but how signals travelled around the network only became clear once Benjamin Franklin went out in a thunderstorm holding a metal key and a kite. In doing so, he performed an experiment that led to the discovery of electricity. This allowed scientists to reconsider how our nerves could be receiving and exchanging signals, eventually inspiring Emil du Bois-Reymond to confirm that electricity is the conduit by which information is exchanged in our neural pathways.

Centuries of research uncovered the highway of cells travelled by electrical signals to govern our actions, but chemical signals are required too. Neurons don't touch one another like a continuous road, but instead are separated by small gaps of water containing charged atoms, or ions. These gaps are known as synapses, which act as a bridge between neurons. Once a signal reaches the end of a neuron, at the axon, compounds called neurotransmitters are released. These cross the bridge and bind to the dendrites – the signal receivers of neighbouring neurons. This causes the next neuron to 'fire' its own electrical signal.

Despite this crossing happening between every neuron, the remarkable structure of the nervous system allows signals to be sent across the body in a fraction of a second. While we now understand much about the broader organisation of nerves and the way they function on a single-cell level, the nervous system continues to hold many mysteries for us to solve.

UNDERSTANDING BIOLOGY

Divisions of the nervous system

Learn the hierarchies of nerve structures that keep our bodies running and safe from harm

When we think of the nervous system, our thoughts immediately go to the brain. The brain is a hive of neuronal activity, with billions of interwoven neurons firing to preserve and recall memories, coordinate thoughts and speech and plan future actions. Along with the spinal cord, the bone-clad parts of our nervous system are called the central nervous system. The majority of our neurons are shielded behind protective fluid and bone, where they receive signals from and dictate to organs around the body. However, the signals sent from the central nervous system must have some means of reaching their target organs. For that they need to connect to nerves that stretch from the central nervous system all the way to the extremities of the body. This second network of nerves is called the peripheral nervous system. Together the central and peripheral form the major divisions of the nervous system.

The peripheral nervous system is responsible for many functions, and as such has numerous subdivisions that specialise in different tasks. The sensory, or afferent division receives signals from the periphery and carries these into the central nervous system. The motor, or efferent division transmits signals for actions outwards from the central nervous system to the peripheral organs and muscles.

These motor functions come in two forms: somatic and autonomic. Somatic functions are perhaps the easiest concept of the nervous system to grasp, as these dictate all of our voluntary actions, such as choosing to pick up a cup or jump on the bed.

Not all somatic motor functions are voluntary, however. Some are automatic, preprogrammed responses built into our bodies that help us cope with danger, known as somatic reflexes. You'll notice such a reflex when you accidentally touch a hot stove, step on a sharp object or something flies towards your eye – your body reacts before you're aware of it. Your hand pulls away, you hop onto the other foot or your eyelid slams shut. This is all the work of somatic reflexes, which can act incredibly quickly as they do not require voluntary input from the brain. Such reflexes can come in different flavours – pulling your hand away from danger is known as a flexor or withdrawal reflex, whereas stepping on a sharp object initiates a crossed-extensor reflex. This latter reflex automatically triggers multiple motor functions: as one leg retracts the other leg simultaneously expands and becomes more stable, preventing us from falling over.

The innate, hard-wired reflex responses of our peripheral nervous system help keep us safe from danger, but they are not the only automatic functions performed by the peripheral nervous system. When actions are not somatic, they are autonomic, which means they operate independently of conscious thought. Such processes include your heartbeat, the churning of food in the digestive tract by contracting muscles and respiration. While our brain can assume control of a few of these processes – think of holding your breath – autonomic functions will continue to operate even when we fall asleep or are unconscious.

The processes that we cannot control, however, are by no means unchanging. Instead the organs under the control of the autonomic nervous system are regulated by a balance between the sympathetic and parasympathetic nervous systems. Depending on stimuli, such as external sensations, these systems increase or decrease activity of our internal organs, helping to ensure our body is always ready to respond to the challenge at hand.

Did you know? Around 80 per cent of the brain's volume is filled with cells

Unlike many peripheral nerves, the nerves of the central nervous system are surrounded by protective bone

Regions of the brain

The brain makes sense of the world around us. It collects information from the five senses, interprets them and returns with instructions for the body's muscles and organs using its three major domains. The cerebrum is the largest part of the brain. It's split into two hemispheres, separated by a fissure that's visible when looking at the brain from above. The left hemisphere controls muscles on the right side of the body, and vice versa.

As well as motor control, the cerebrum is responsible for interpreting touch, vision and sound; controlling our sense of reasoning and determining our emotions. The cerebellum, nestled under the cerebrum, coordinates posture, balance and other muscle movements. Finally, the brainstem acts as a conduit, connecting the other regions of the brain to the spinal cord.

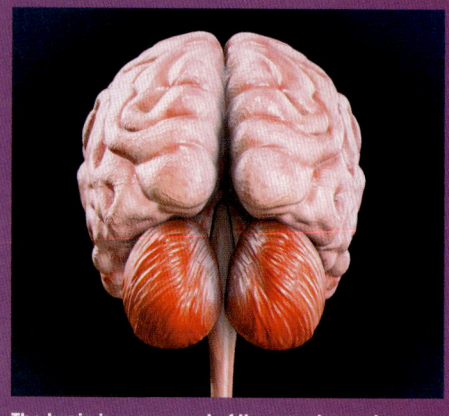

The brain is composed of three major regions; the cerebrum is the largest

HUMAN BIOLOGY

UNRAVELLING THE
NERVOUS SYSTEM

The reflex arc
Your nervous system rescues you from imminent harm using hard-wired reflex actions

PAIN

RESPONSE

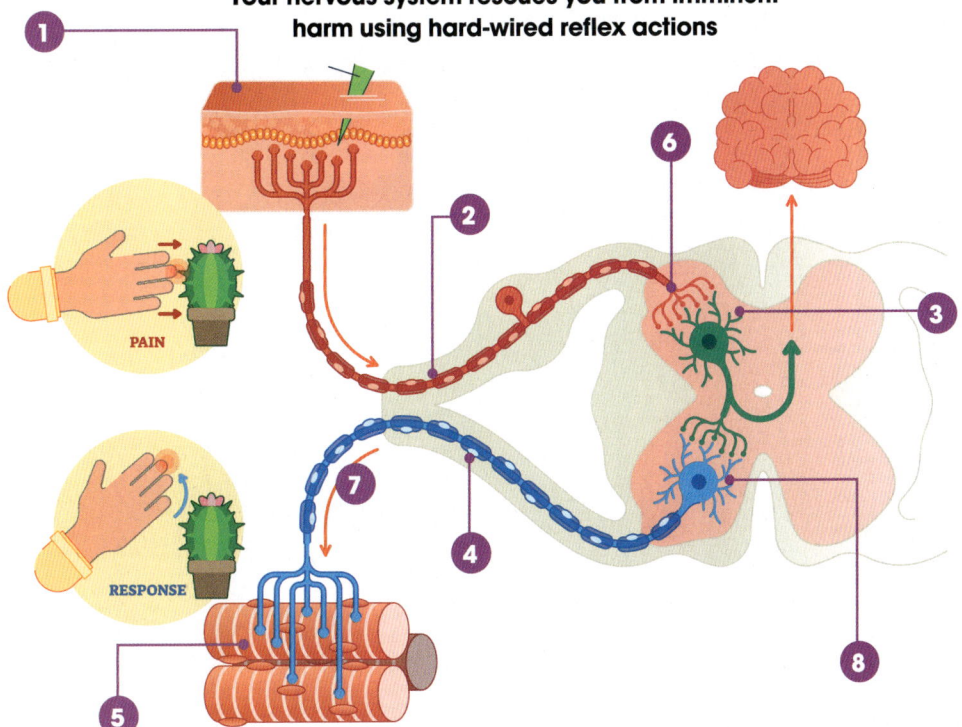

Types of neurons
Not all neurons play the same role

Unipolar
A useful way of defining neurons is by the number of processes, or branches, that extend from the cell body. A unipolar neuron has just one axon branch.

Bipolar
These neurons have one dendrite and one axon extending in opposite directions. They are found up your nose and behind your eyes.

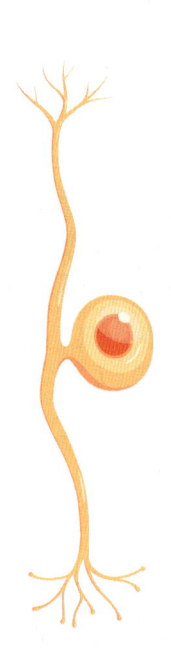

Psuedounipolar
These have one branch that extends from the cell body and splits into two branches. These neurons help connect the peripheral to the central nervous system.

Multipolar
Our most common type of neuron, which is numerous in the central nervous system. These possess many dendrites and a single axon.

1 Detection
When your cells are at risk, such as when you accidentally touch a hot stove, receptors in your skin initiate a nerve impulse.

2 Sensor neuron
The nerve impulse travels from the dendrites of the sensory neuron towards the spinal cord.

3 Interneuron
Located in the spinal cord, the interneuron acts as a relay between the sensor and motor neurons, passing the impulse onto the latter.

4 Motor neuron
The nerve impulse travels away from the spinal cord and towards the target muscle receptors via the motor neuron.

5 Motor function
The signal forces muscles to contract, pulling your hand away from the heat source automatically without any conscious thought needed on your part.

6 Close neighbours
The nerve impulse reaches connected neurons by 'jumping' across the synapse, a small gap that lies between them.

7 One-way traffic
The nerve impulse travels in one direction, starting at the sensory neuron and terminating when it gets to the end of the motor neuron.

8 Polysynaptic reflex
Reflexes can trigger either simple or complex responses. Polysynaptic reflexes can send signals to many muscles, allowing us to simultaneously contract them during a reflex.

43

Relaxing or taxing

How we feel can affect our body in many ways

Sympathetic nervous system

When we feel threatened, our instinctual reaction is known as the fight-or-flight stress response. The response induces a suite of physiological actions to help us cope with the challenge.

Dilated pupils
Adrenaline causes the pupils to widen, allowing the eyes to absorb more information from their surroundings.

Inhibited salivary glands
Increased noradrenaline depresses the activity of saliva-producing acinar cells, helping to slow the digestive tract.

Expanded airways
Adrenaline expands the airways lining the lungs, allowing the increased blood supply to more readily oxygenate.

Increased heart rate
The heartbeat quickens following exposure to adrenaline. This improves the blood supply to muscles and the brain.

Spinal connections
Organs connect to the sympathetic nerves via distinct ganglia that emerge from different regions of the spine.

Acetylcholine
Much of the parasympathetic nervous system involves muscarinic receptors, which are activated by the molecule acetylcholine.

Parasympathetic nervous system

When we're free from periods of stress, our body responds by reverting to a more balanced state of rest and digest.

Pupil constriction
Stimulated muscarinic receptors in the eye cause contraction of the iris' sphincter muscle and the ciliary muscle, improving near vision.

Stimulated saliva secretion
Parasympathetic stimulation causes the secretion of saliva components, including water, amylase and potassium ions.

Constricted airways
The large tubes of the lungs, known as bronchi, constrict and stimulate secretions. These processes help protect the lungs from airborne irritants.

Decreased heart rate
Parasympathetic inputs reduce the velocity of electrical conduction through the atrioventricular node of the heart, helping to slow contractions and reduce heart rate.

Did you know?
Cortisol reduces inflammation, helping keep nerves healthy

C1 C2 C3 C4 C5 C6 C7 C8 T1

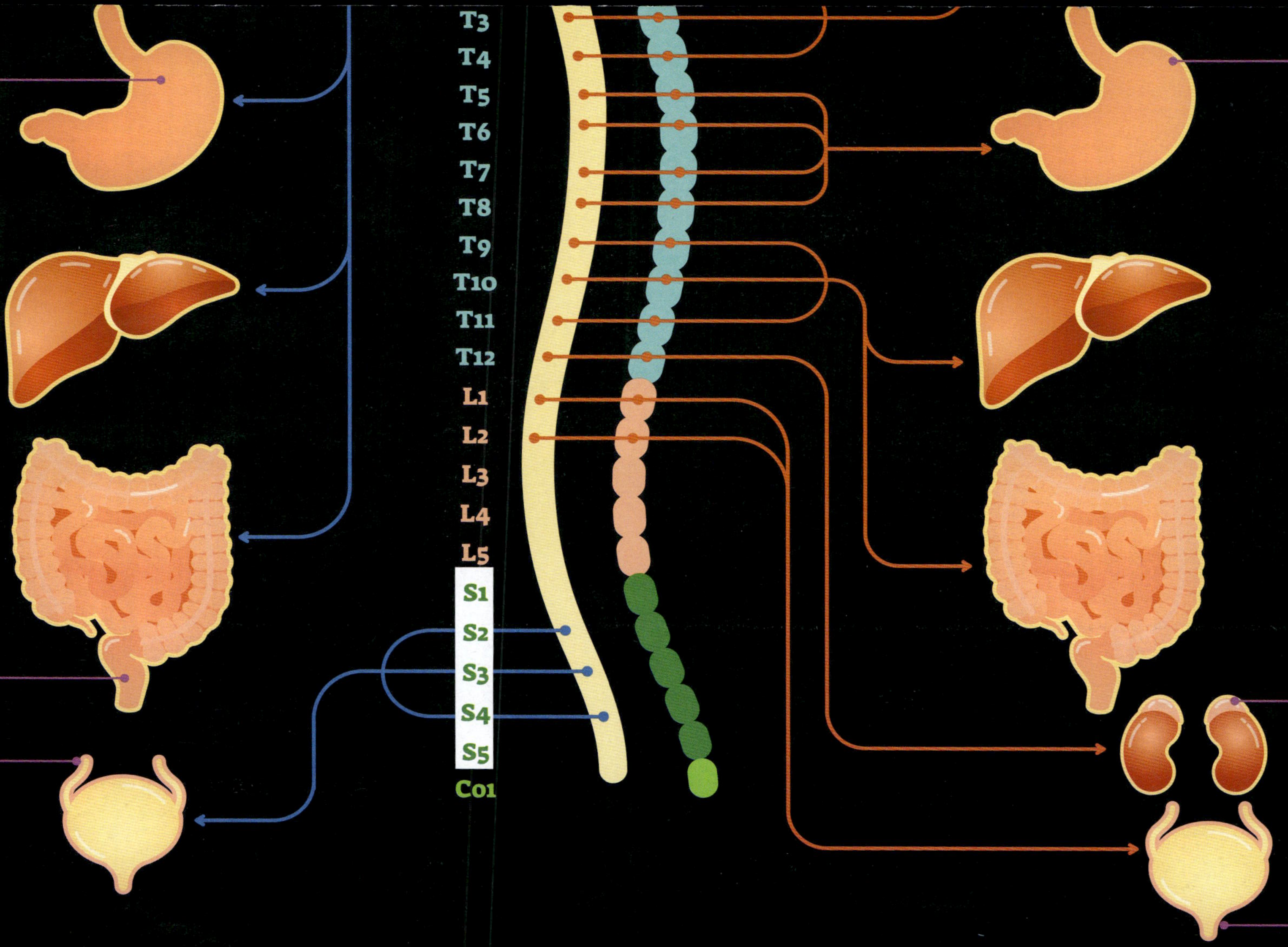

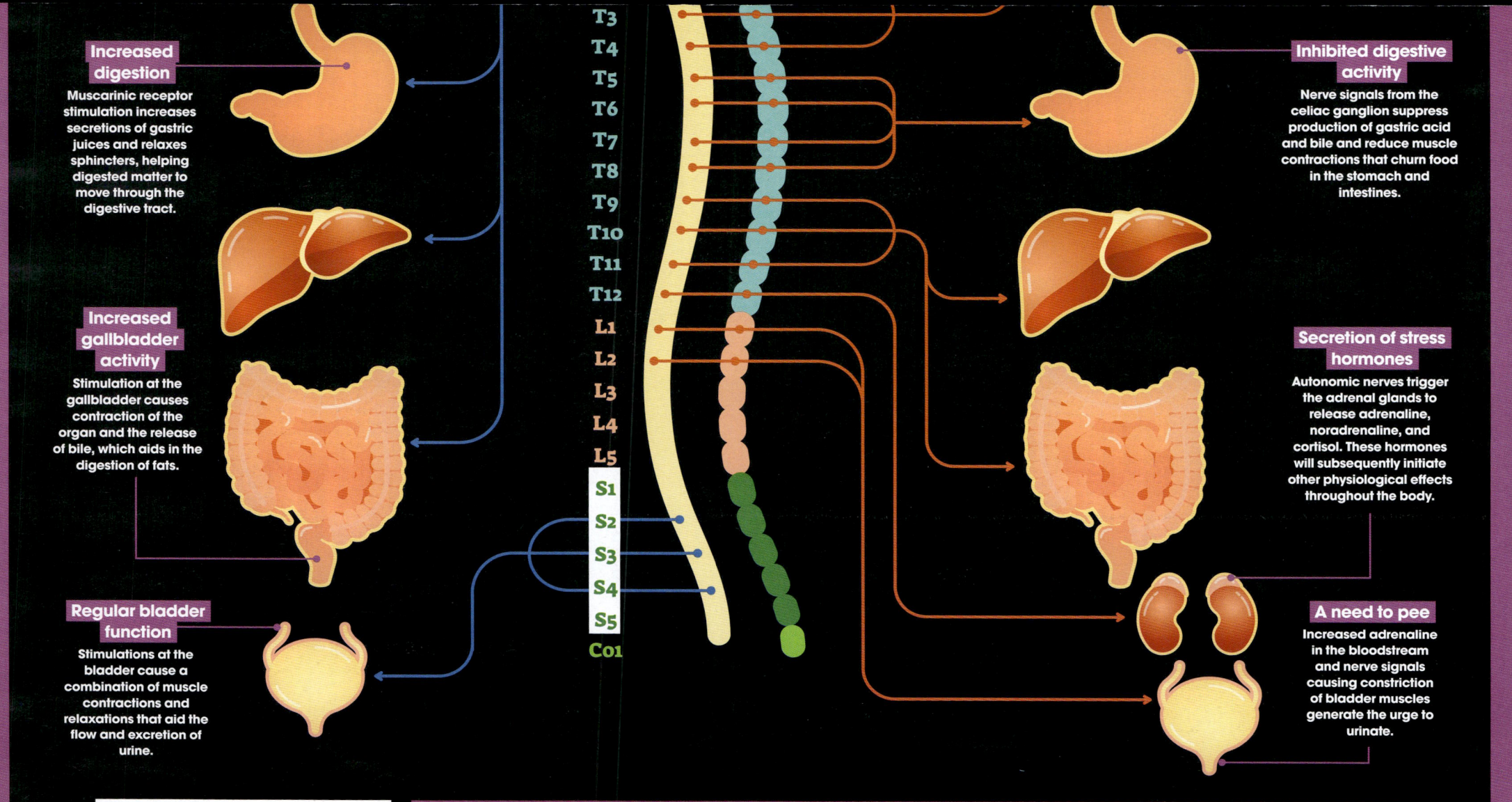

Inhibited digestive activity
Nerve signals from the celiac ganglion suppress production of gastric acid and bile and reduce muscle contractions that churn food in the stomach and intestines.

Secretion of stress hormones
Autonomic nerves trigger the adrenal glands to release adrenaline, noradrenaline, and cortisol. These hormones will subsequently initiate other physiological effects throughout the body.

A need to pee
Increased adrenaline in the bloodstream and nerve signals causing constriction of bladder muscles generate the urge to urinate.

Increased digestion
Muscarinic receptor stimulation increases secretions of gastric juices and relaxes sphincters, helping digested matter to move through the digestive tract.

Increased gallbladder activity
Stimulation at the gallbladder causes contraction of the organ and the release of bile, which aids in the digestion of fats.

Regular bladder function
Stimulations at the bladder cause a combination of muscle contractions and relaxations that aid the flow and excretion of urine.

T3
T4
T5
T6
T7
T8
T9
T10
T11
T12
L1
L2
L3
L4
L5
S1
S2
S3
S4
S5
Co1

A bear or a boss?
The fight-or-flight sympathetic stress response evolved to help humans survive encounters with predators. For much of our hunter-gatherer existence, as we explored the wilderness, discovered new environments and spread across the globe, we encountered all manner of dangerous fauna. For some of these encounters, quick reaction times would have been vital, helping those with an attuned response survive and pass on their genes to the next generation. Today most of us have little to fear from a hungry panther or a territorial grizzly bear, but the stress response is still routinely triggered, just by other means. Now many of us encounter the same intrinsic response when we run into an angry teacher in the corridor or when we have to present unexpectedly to our employer's CEO at the annual meeting.

Today fight or flight often triggers when we're socially anxious

Relaxation techniques such as yoga can help tackle overstimulation of the sympathetic nervous system

© Getty

UNDERSTANDING BIOLOGY

Diseases and Damage

As the connective network coordinating our body, damage to the nervous system can be debilitating

5 ways cortisol affects us

1 All over
Cortisol is the main stress-response agent in the body. Adrenaline and noradrenaline affect multiple organs, but cortisol is ubiquitous – most cells in the body possess cortisol receptors.

2 Sounding the alarm
The hypothalamus is a small region of the brain that monitors and regulates the stress response. During periods of stress, the hypothalamus signals the pea-sized pituitary gland to stimulate cortisol release.

3 Give me fuel
During the fight-or-flight response the body may need to act quickly, and for that it needs energy. Cortisol provides this fuel by stimulating the release of glucose.

4 Priorities
Non-vital organs for survival during periods of stress, such as the reproductive organs and immune system, are suppressed by cortisol, ensuring energy goes where it's needed most.

5 Overstressed
Cortisol aids during periods of stress and during daily function, but chronic stress overstimulates cortisol release. This risks a multitude of health problems, including anxiety and heart disease.

Pinched nerve

Excessive pressure from nearby tissue, including spinal discs, can 'pinch' nerves and interfere with their signalling. This results in pain, tingles, weakness or numbness as the nerve misfires

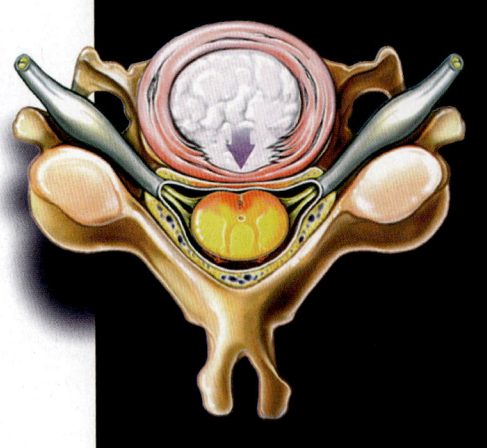

Disc bulge
Spinal discs rest between the vertebrae of the spine, where they act as shock absorbers. They are formed of a softer centre and tougher outer band. As we age, cartilage in the disc dries and stiffens and the outer layer bulges.

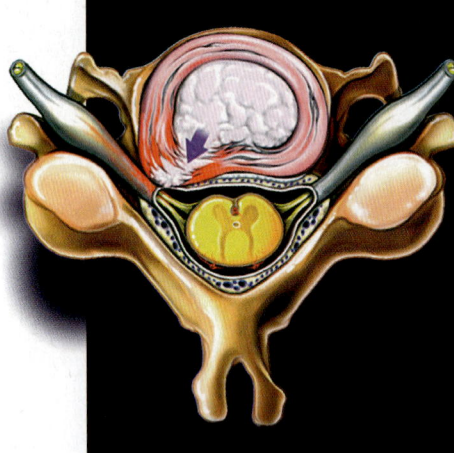

lateral disc herniation
When the outer layer of the spinal disc ruptures, it's known as a herniation, which allows the softer centre of the disc to ooze out. In laterally herniated discs, this places pressure on the nerve root of the spinal cord.

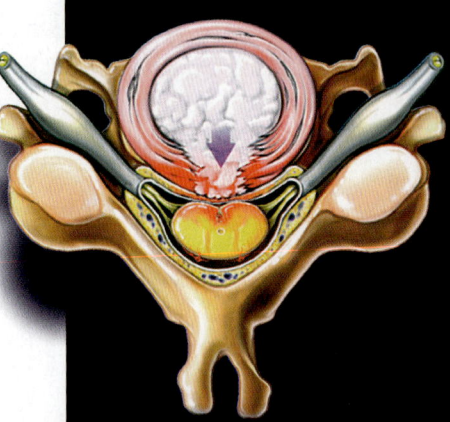

central disc herniation
Central herniation can place excessive pressure on the spinal cord, causing severe compression in some cases. This can result in loss of motor function of the legs, affecting balance and strength when standing and walking.

Spinal surgery can be used to treat herniated discs by alleviating pressure on the spinal cord

Bell's Palsy

Nerves spread out from the central nervous system to various organs and tissues, with each powering specific functions. Cranial nerve VII is known as the facial nerve, as it controls many of the muscles on our face, including blinking and smiling. When this nerve is inflamed, damaged or disrupted, Bell's palsy can occur, with the facial muscles becoming weakened or paralysed. This typically affects only one side of the face, causing drooping of the mouth on one side and a loss of control of an eyelid, giving the affected side a slack appearance. The full symptoms of Bell's palsy are often temporary, with some or total recovery of the affected areas occurring within six months. While it's not always clear what causes the cranial nerve to swell and Bell's palsy to occur, scientists believe that a recurring viral infection of the nervous system elicits an immune response that triggers the nerve damage and causes the symptoms.

Bell's palsy causes paralysis in part of the face due to swelling of cranial nerve VII

HUMAN BIOLOGY

Multiple sclerosis

Neurons are the agents of signalling in our bodies, but they don't work alone. Axons, which carry signals away from the neuron's cell body, are coated in a sheaf of myelin. Myelin sheaves are produced in the central nervous system by cells called oligodendrocytes, enabling myelin's function of protecting and facilitating nerve conductivity. In multiple sclerosis, a severe abnormal immune response within the central nervous system strips away the protective myelin and subsequently causes lots of nerve scarring, or sclerosis, which gives the disease its name. Research efforts are underway to treat the disease by encouraging myelin regeneration.

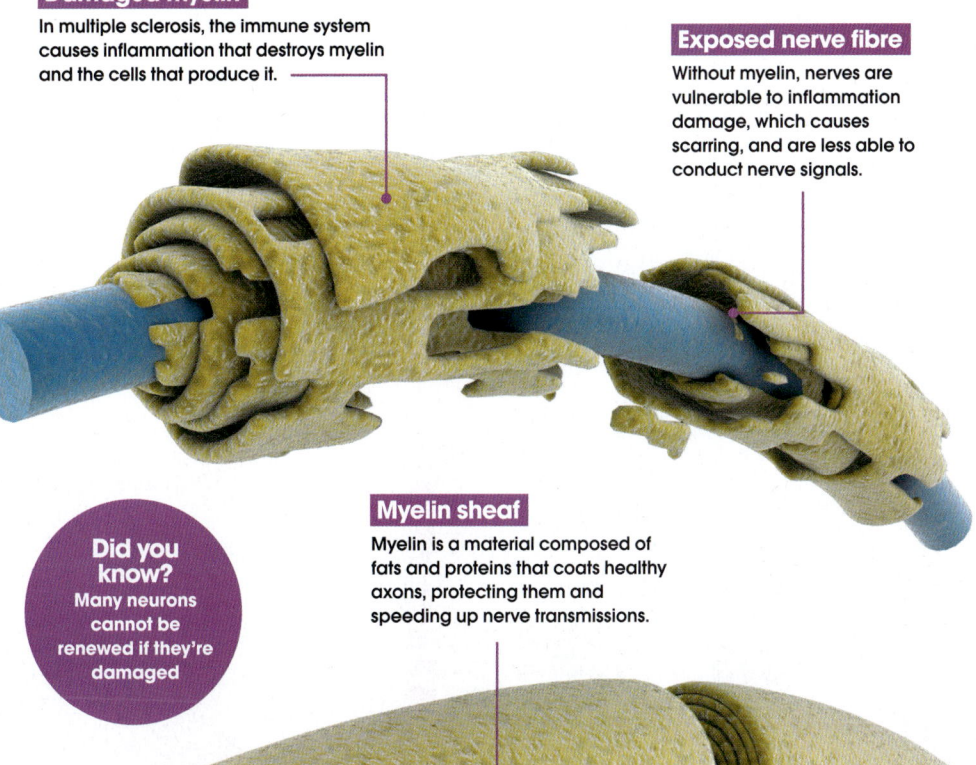

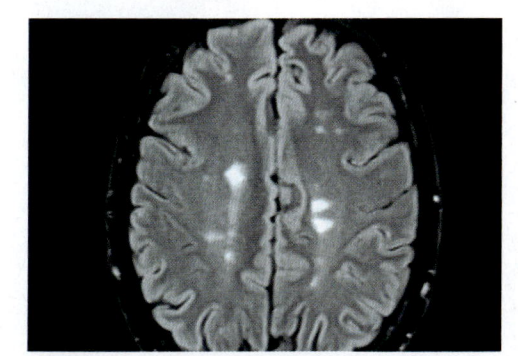

Damaged myelin
In multiple sclerosis, the immune system causes inflammation that destroys myelin and the cells that produce it.

Exposed nerve fibre
Without myelin, nerves are vulnerable to inflammation damage, which causes scarring, and are less able to conduct nerve signals.

Did you know? Many neurons cannot be renewed if they're damaged

Myelin sheaf
Myelin is a material composed of fats and proteins that coats healthy axons, protecting them and speeding up nerve transmissions.

Multiple sclerosis damages both heavily myelinated white matter and grey matter at the surface of the brain

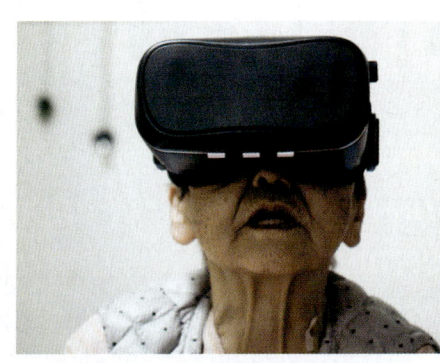

Research is in progress to use virtual reality as a potential diagnostic tool for neurological disorders

Peripheral neuropathy from diabetes can result in a loss of sensation in the feet

Peripheral neuropathy

While several degenerative disorders primarily or exclusively impact the central nervous system, there are a collection of diseases that instead impact the peripheral nervous system, called peripheral neuropathies. As the impacted region is the peripheral nervous system, such neuropathies lead to a loss of sensation and regulatory control of extremities. These include a loss of coordination and feeling in fingers and toes and a lack of balance. The causes of peripheral neuropathy are yet to be fully made clear, but scientists have determined diabetes, which causes protracted periods of high blood sugar, as one of the primary causes.

Early symptoms of Parkinson's include tremors in the hands

Parkinson's

The basal ganglia are found deep within the brain, in an area responsible for controlling movement. These nerves produce a compound known as dopamine, which is important in coordinating numerous functions, including executive functions and motor control. Although the cause is not yet clear, the basal ganglia can become impaired and begin to die. The result of this is Parkinson's disease, as the loss of dopamine gradually hampers key functions such as walking, talking and memory recall. These effects are compounded by the loss of nerves responsible for producing norepinephrine, a key compound in the sympathetic nervous system needed to regulate heart rate and blood pressure.

UNDERSTANDING BIOLOGY

How hormones control your body

Just as the nervous system sends information around the body via electrical impulses, the endocrine system provides another messaging service that has complete control of your body

HUMAN BIOLOGY

HOW HORMONES CONTROL YOUR BODY

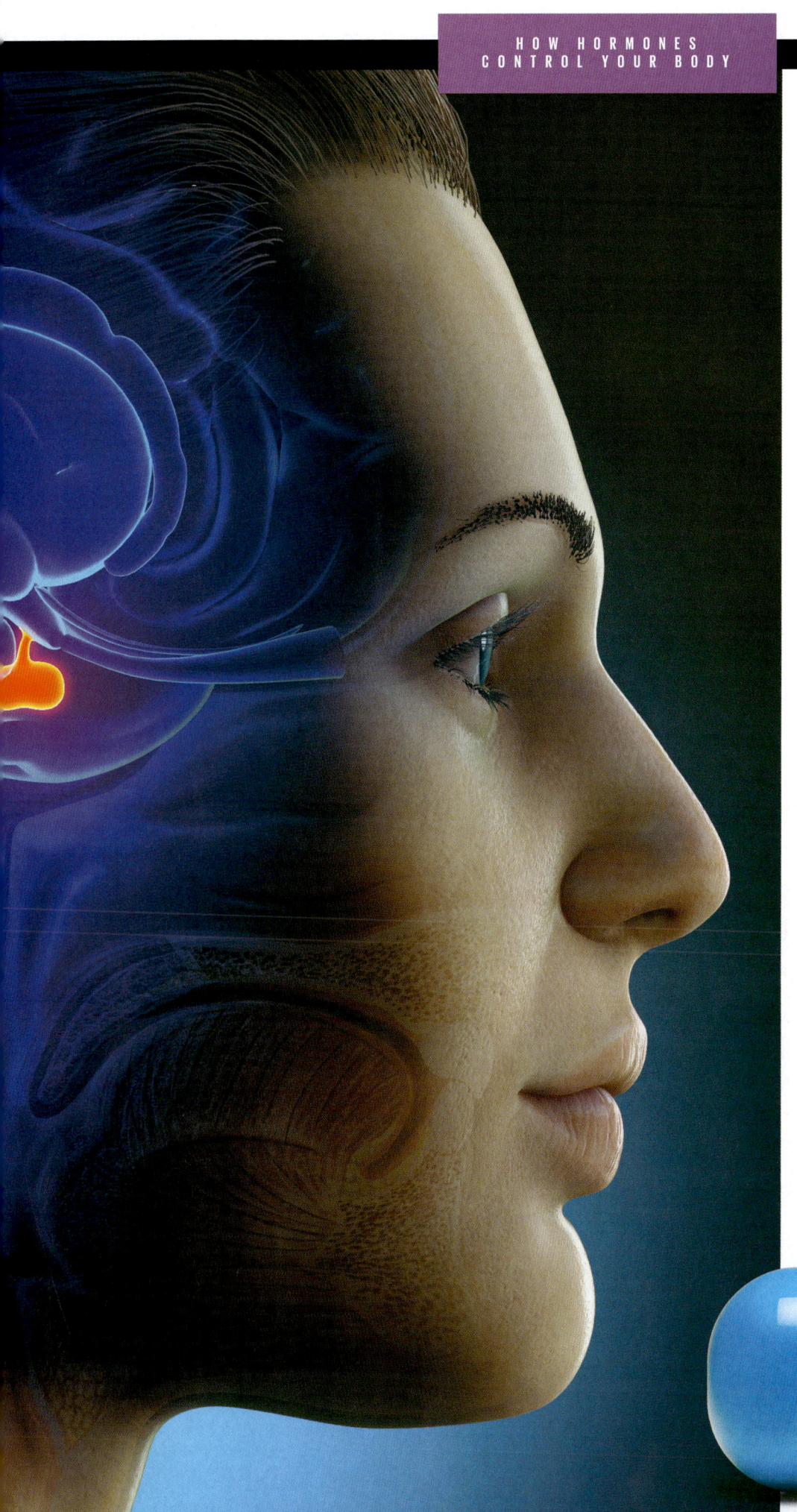

If you've ever wondered why you never seem to get any taller while your friends are regularly growing out of their clothes, why your skin always breaks out in spots when you want to look your best and why you're always hungry no matter how much you eat, you need look no further than your hormones. But what exactly are hormones, and why do they seem intent on interfering in almost every aspect of your life?

In simple terms, hormones are chemical molecules that act like an internal postal service. These specialised proteins carry vital messages around the body via the bloodstream – thanks to the circulatory system and a strong, beating heart – to their target cells, where they give out clear instructions. Different types of hormones will give instructions to entirely different organs and tissues within the body. This ingenious system begins before your birth and continues to make appropriate changes to your body throughout your entire life, allowing you to develop, grow, thrive and survive.

The endocrine system plays an important role in the transition from foetus to independent newborn baby. Endocrine cells have started to disperse within a few weeks of gestation, and by 13 weeks have developed a hypothalamus and a pituitary gland. The foetus takes hormones from the parent and the placenta acts as a temporary endocrine organ, sharing nutrients and messages between the two. This is expelled as the baby is pushed out. The hormones help the foetus to develop and grow in utero, giving it the best chance to survive the birthing process. As soon as the baby is born, the endocrine system controls every aspect of its life, from growth and strength to temperature and mood. However, it's when reaching puberty – usually around the age of 12 – that most people become aware of the importance of hormones.

Many women overcome menopausal side effects by undertaking hormone replacement therapy

HRT

49

UNDERSTANDING BIOLOGY

It's during puberty that our bodies develop into their peak physical state and complete sexual maturity. The endocrine system instigates some dramatic bodily changes that can be very confusing and frightening for any unprepared teenager. As adolescence begins, the hypothalamus and pituitary gland stimulate the gonads (ovaries or testes) to produce a variety of hormones that prompt new sexual characteristics. Oestrogen and progesterone increase in girls, causing the breasts to enlarge and the menstrual cycle to begin. Meanwhile, an increase in testosterone in boys initiates a deepening of the voice, enlargement of the sexual organs and an increase in facial hair. Temperament also alters, and often not for the better. In medieval times, before there was any understanding of hormonal flux, previously angelic children were sometimes thought to have been possessed by the devil when they reached adolescence.

But any system can malfunction, and the endocrine system is no different. An imbalance of hormones in a young child or adolescent can bring about a multitude of disorders. Precocious puberty occurs when a child begins the transition into sexual maturity at a very early age. The hypothalamus sends signals to release sexual hormones too soon, causing little girls to start their periods before the age of eight. This can sometimes be brought about by a tumour growing on one of the key endocrine glands. Meanwhile, a lack of the growth hormone can lead to dwarfism, and a deficiency in the hormone insulin can lead to childhood diabetes. It's vitally important that parents maintain a watchful eye over their children to ensure that any endocrine issues are dealt with swiftly, since early treatment of a hormonal imbalance can prevent long-term health problems.

The endocrine system plays a vital role in adulthood, too. Hormones such as insulin and glucagon support the regulation of metabolic health, since they balance blood sugar levels. As our lives become more complicated with financial commitments and employment obligations, adults need to be capable of remaining calm under pressure in order to stay healthy. The adrenal glands release cortisol and adrenaline to combat spiking stress levels and maintain stable blood pressure. The T3 hormone released from the thyroid gland works alongside calcitonin to maintain healthy strong bones and proper growth.

A crucial role for the endocrine system in an adult's life concerns reproductive hormones. Oestrogen and progesterone, starting in puberty, continue to be extraordinarily important, as they not only synchronise the menstrual cycle, but also support cardiovascular

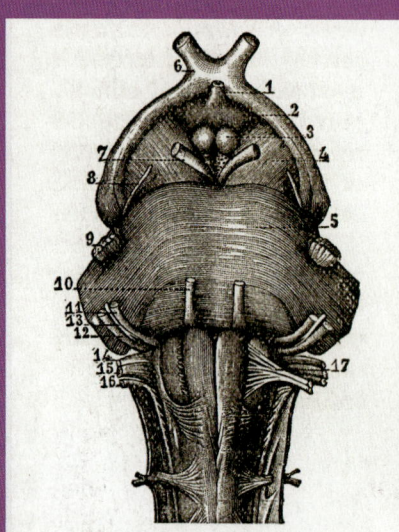

The infundibulum of the pituitary gland is included in this Victorian anatomical illustration of the cerebral medulla

Hashimoto's thyroiditis occurs when antibodies attack the thyroid gland, causing an unhealthy hormone imbalance

Thanks to an excess of growth hormone, Robert Wadlow reached a height of 271.8 metres

'Mucus Store'

Scientists have come a long way in their understanding of the pituitary gland and its many functions, but although it's only relatively recently that we have fully understood its importance, we've been aware of its presence for centuries. As long ago as 150 CE, Greek physician and philosopher Galen identified the gland and, noting its ability to secrete substances, suggested that its purpose was to help drain phlegm from the brain. By the early 18th century, doctors had discovered various anatomical elements of the gland, such as the hypothalamo-hypophyseal axis linking the pituitary gland to the hypothalamus – although at that time they were unaware how important this connection was. Various pituitary-related diseases were discovered throughout the 18th century, and by the mid-19th century Dr Martin Rathke described its formation. As time progressed, scientists discovered how vital the pituitary gland is to our continuing survival and how it is manipulated by the hypothalamus.

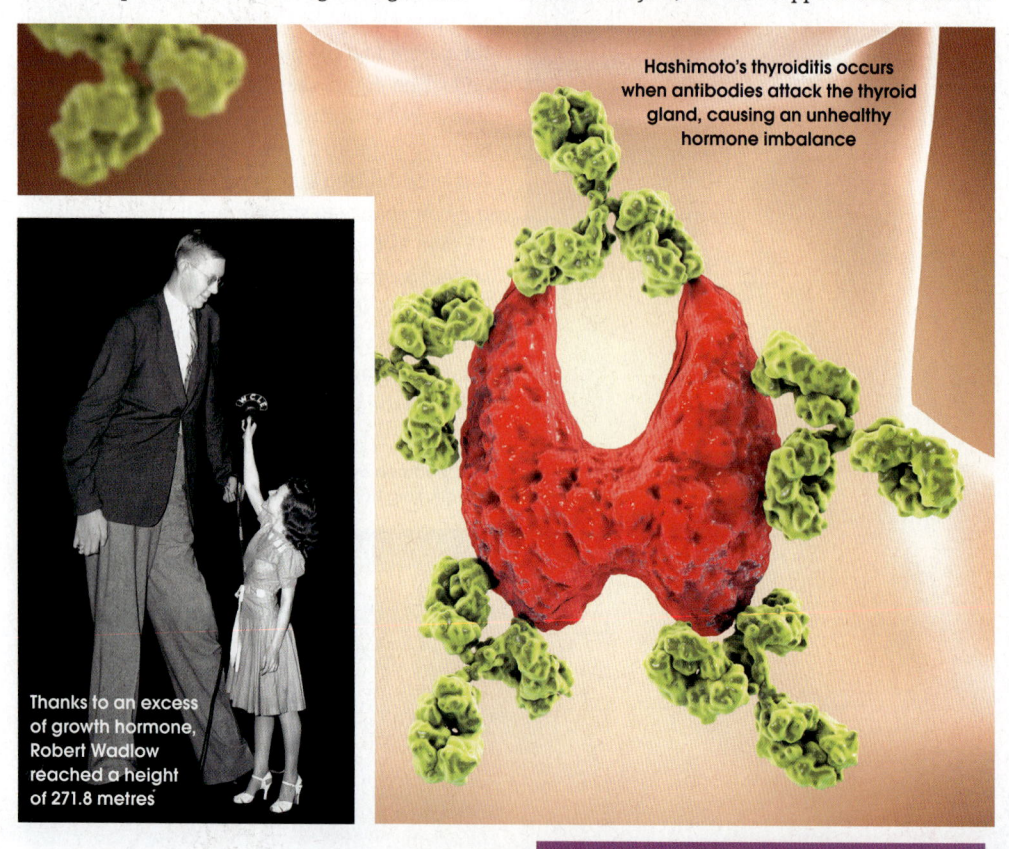

health, bone health and brain function. The same can be said of the male hormone, testosterone, as it also supports bone health and muscle formation alongside general reproductive wellbeing.

When a person becomes pregnant, the endocrine system needs to work doubly hard as it is supporting two individuals and must finely balance the needs of the parent with the requirements of a developing foetus, while also accommodating the parent's physical bodily changes. Many of the common complaints that arise during pregnancy, such as swollen ankles, heartburn, raised blood pressure and constipation, are all dealt with via hormones.

As we slide into old age, the endocrine system instigates more hormonal changes, which in turn causes a multitude of alterations and modifications to our bodies. Oestrogen levels lower, initiating menopause, a state where the body can no longer reproduce. The growth hormone diminishes, as does testosterone, so both males and females become more clinically vulnerable. Bone density and muscle mass decrease and the immune system weakens. Glands within the endocrine system take longer to produce hormones and often secrete far less than when the body was younger, while the hormones themselves take far longer to break down. Such changes negatively affect the body's functions, creating distressing characteristics in the elderly such as memory loss and disorders like osteoporosis.

Without the endocrine system creating and secreting the many hormones that enter our bloodstream and ordering our organs and tissues to behave in a particular way, we wouldn't be the fascinating and complex creatures that we are. So next time someone asks you why you're being moody or why you need that extra hour in bed, you can answer truthfully... it's my hormones.

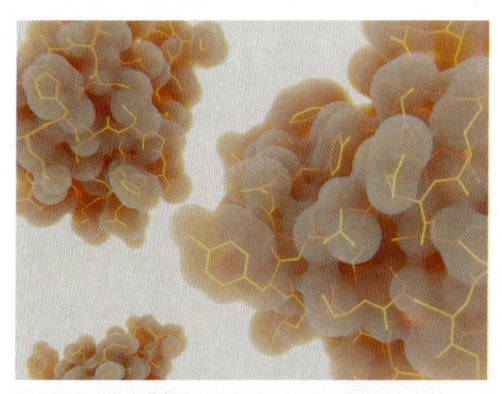

Insulin regulates the metabolism of carbohydrates, proteins and fats by promoting the uptake of glucose

The Master Gland

The pituitary gland is also known as the master gland since it oversees all other hormonal instructions. Hormones from the hypothalamus enter the hypophyseal portal, which connects the hypothalamus to the pituitary gland. A network of capillaries carry the hormones to the anterior lobe of the pituitary, where they start or stop the production of pituitary hormones.

These hormones stimulate various endocrine glands that control a multitude of bodily functions. The thyroid-stimulating hormone, for example, instigates parts of the body's metabolism such as temperature control, while the human growth hormone sends instructions to the skeleton, the liver and the muscles. Two hormones, oxytocin and vasopressin, are sent to the posterior pituitary lobe.

Vasopressin, or antidiuretic hormone (ADH), controls water loss in the body, while oxytocin is vitally important during childbirth. The pituitary gland works in conjunction with the hypothalamus to help our bodies deal with situations such as stressful events.

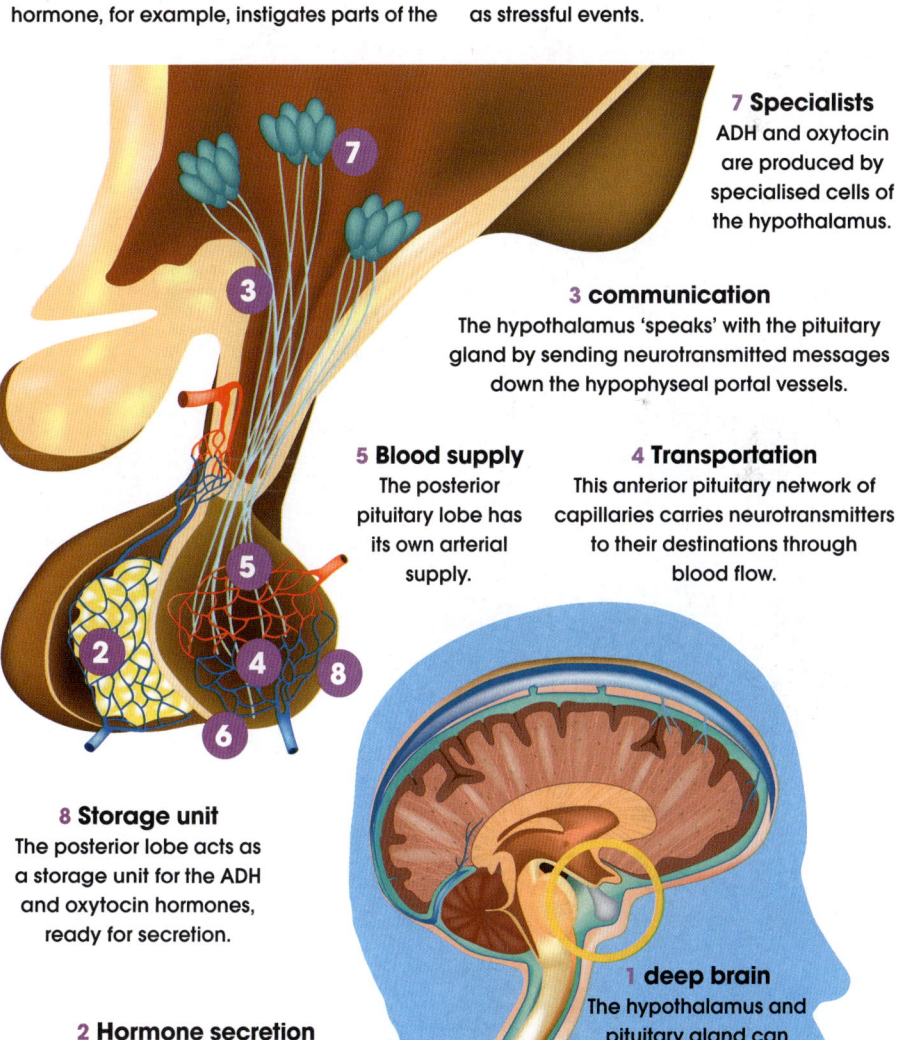

7 Specialists
ADH and oxytocin are produced by specialised cells of the hypothalamus.

3 communication
The hypothalamus 'speaks' with the pituitary gland by sending neurotransmitted messages down the hypophyseal portal vessels.

5 Blood supply
The posterior pituitary lobe has its own arterial supply.

4 Transportation
This anterior pituitary network of capillaries carries neurotransmitters to their destinations through blood flow.

8 Storage unit
The posterior lobe acts as a storage unit for the ADH and oxytocin hormones, ready for secretion.

2 Hormone secretion
The adenohypophysis is the largest part of the anterior lobe and is responsible for hormone secretion.

1 deep brain
The hypothalamus and pituitary gland can be found in the diencephalon area of the brain.

UNDERSTANDING BIOLOGY

Glands of the Endocrine System

Eight major glands make up the endocrine system, and each one secretes a different hormone

Targeting Cells

How do hormones know which cells to bind with and how do they get there?

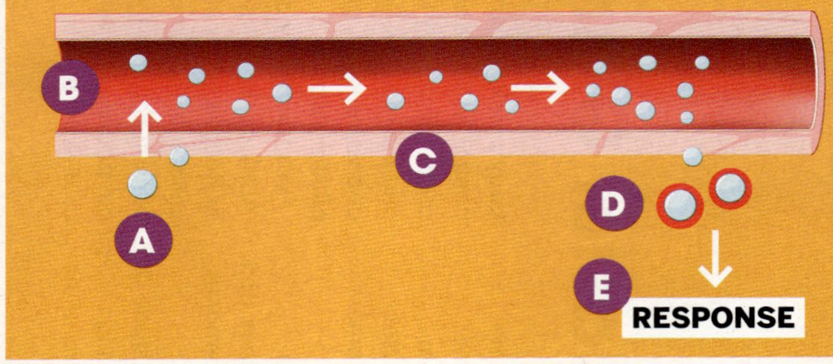

A The Endocrine Gland
The hypothalamus instigates each ductless endocrine gland to produce a particular hormone.

B The Body's Highway
The hormone is secreted directly into the circulatory system via the bloodstream, which acts as a road network.

C Non-Target Cells
Without an appropriate receptor, the hormone will ignore a cell and keep moving along the bloodstream.

D The Target Cell
Receptors on the cell's surface act like a beacon, informing hormones which organs they're intended for.

E The Binding Process
The hormone binds with the target cell, triggering a series of reactions that will alter the behaviour of that cell.

Did you know? Too much alcohol can decrease testosterone levels

Endocrine or Exocrine

While both the endocrine and exocrine systems produce and secrete hormones, the difference between them lies in the way they distribute hormones around the body. Unlike the endocrine system, the exocrine system doesn't rely on the bloodstream to transport hormones. It works outside of the bloodstream, planting the hormones directly onto an epithelial surface. This might be the outer layer of the skin – known as the epidermis – the lining of your intestines or the lining of your respiratory tract. The exocrine system achieves this by delivering hormones via a duct. These ducts can come in many different shapes and sizes, from simple to tubular, and can be individual or clumped together.

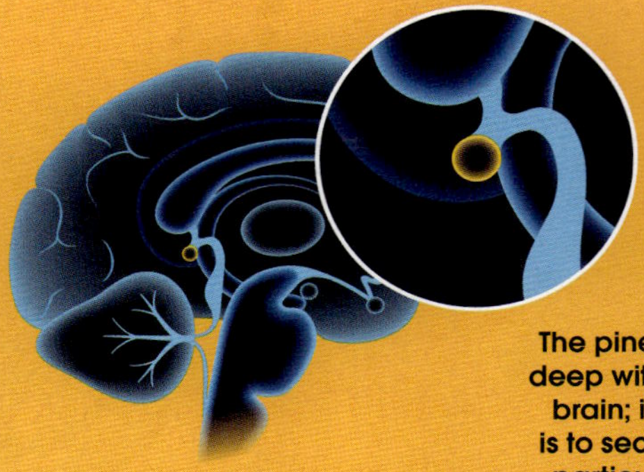

2 Pineal
The pineal gland is located deep within the centre of the brain; its primary function is to secrete melatonin. This particular hormone helps you get to sleep easily and regulates a healthy circadian rhythm.

1 The Pituitary Gland

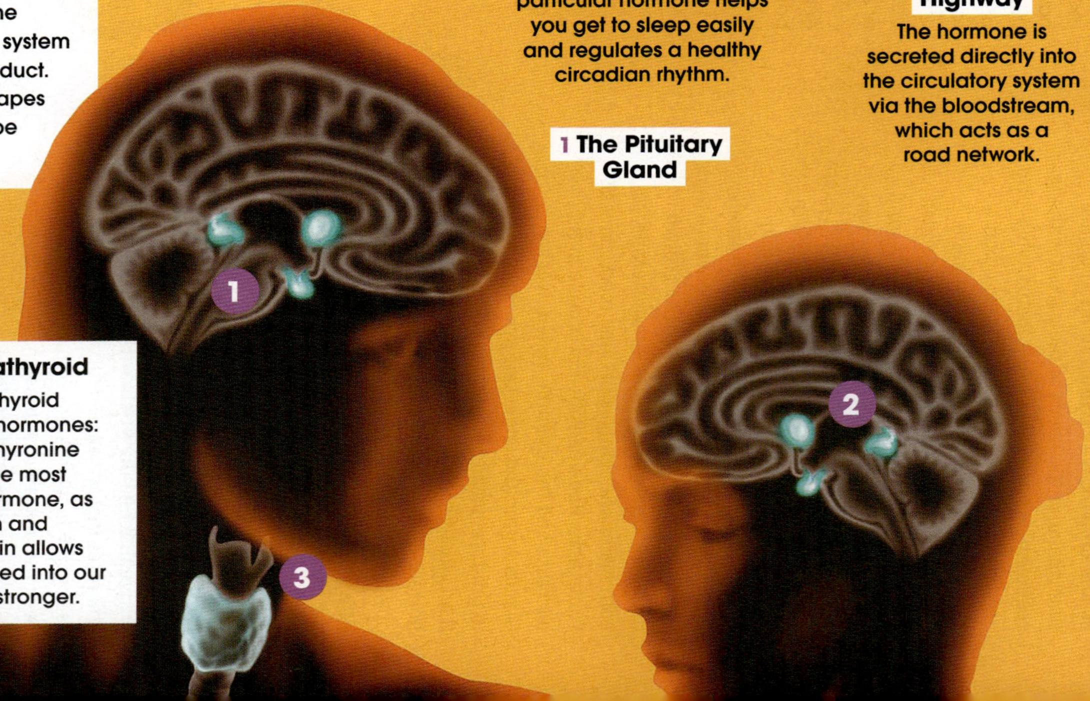

3 Thyroid and Parathyroid
The thyroid and parathyroid glands secrete three hormones: thyroxine (T4), triiodothyronine (T3) and calcitonin. The most important is the T3 hormone, as it controls your growth and metabolism. Calcitonin allows calcium to be absorbed into our bones, making them stronger.

HUMAN BIOLOGY

HOW HORMONES CONTROL YOUR BODY

4 Thymus
Various hormones are made here: the thymopoietin, thymosin and thymulin help in the production of T-cells, while the thymic humoral factor supports the immune system. When enough T-cells have been produced, the gland begins to wither and is replaced with fat.

8 Ovaries
The ovaries create and secrete the hormones oestrogen and progesterone in women, which help start the menstrual cycle and develop breasts at puberty. These hormones are also vitally important during pregnancy.

6 Adrenal
Adrenal glands, which are found at the tops of your kidneys, make vitally important hormones including epinephrine, cortisol, noradrenaline and aldosterone. These affect many bodily functions, including your oxygen intake, blood pressure and stress reactions. They also produce reproductive hormones.

7 Testes
The testes make the hormone testosterone in men, which instigates the growth of facial and body hair, develops a sex drive and helps in the creation of sperm.

5 The Islets of Langerhans
Islets of Langerhans cells create and release insulin, which acts to lower the blood sugar level, and glucagon, which raises blood sugar levels. In this way, glucose levels are maintained within the body.

53

UNDERSTANDING BIOLOGY

Imbalances and disorders

The endocrine system plays a vital role in regulating biological functions. Any imbalance of hormones can have catastrophic results, leading to a variety of disorders

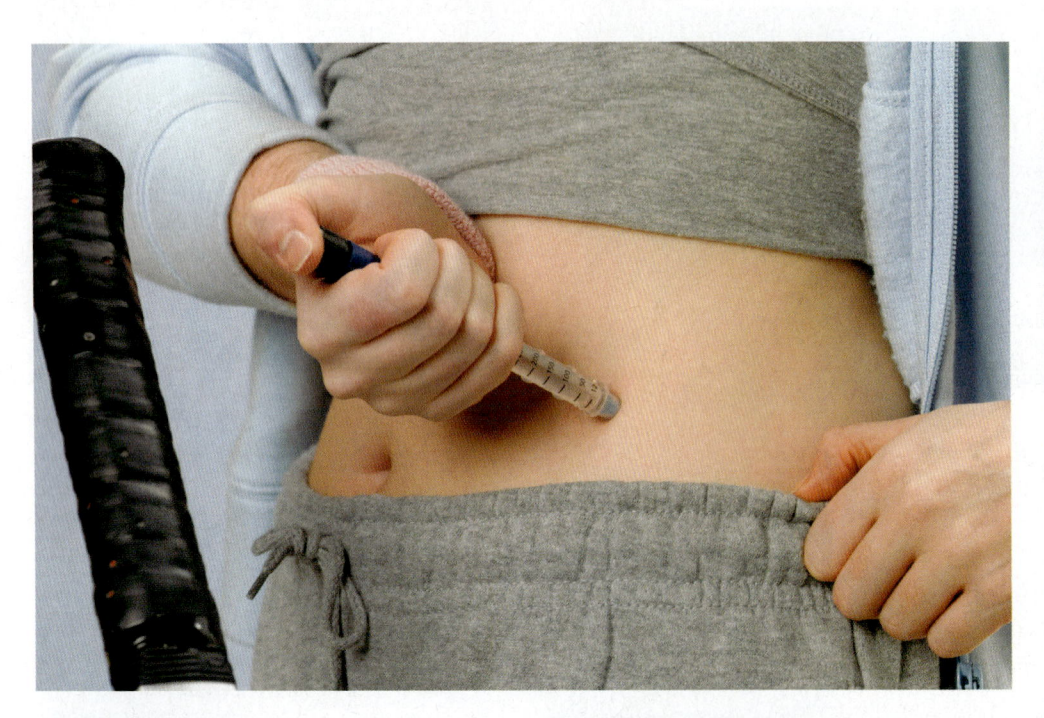

Diabetes

Diabetes occurs when the body is unable to produce the hormone insulin or use it effectively. This can lead to dangerously high blood sugar levels. Diabetes can occur in two forms: type 1 is an autoimmune condition that tends to develop in childhood or teen years. The body's immune system attacks cells in the pancreas, stopping the production of insulin. With type 2 diabetes, the body becomes resistant to insulin, causing a spike in blood sugar levels. Unlike type 1 diabetes, type 2 tends to occur in adults and can sometimes be brought about by eating too many sugary foods. Both type 1 and type 2 can be caused by a combination of genetic and environmental factors. Symptoms can include feeling exceptionally thirsty, having blurry vision, needing to urinate more frequently and feeling exhausted all the time. Long-term complications of diabetes can include cardiovascular disease and kidney damage.

Pituitary Gigantism

This occurs when the pituitary gland produces far too much growth hormone in children before their bone growth plates fuse together. Muscle and organ size and height increase exponentially, causing the child to become abnormally big for their age. It can also cause many other symptoms, including excessive sweating and double vision.

Addison's Disease

This occurs when the adrenal glands fail to produce the required amounts of the hormones cortisol and aldosterone. For this reason it is also known as primary adrenal insufficiency. The symptoms include low blood pressure, muscle weakness and overwhelming tiredness. Addison's disease can be caused by any damage to the adrenal glands.

A healthy parent can give birth to a child with dwarfism

Polycystic Ovary Syndrome

Scientists aren't sure of the exact cause of polycystic ovary syndrome, but they have noticed a correlation between women who suffer with the disease and an increase in the hormone insulin. Some women create more insulin than is necessary because their bodies are resistant to it. This instigates an increase in another hormone, testosterone. When this occurs, the patient may suffer from excessive facial hair and weight gain, and may also find it very difficult to fall pregnant. This is due to infrequent ovulation cycles, and sometimes the body can stop ovulating altogether.

HUMAN BIOLOGY

HOW HORMONES CONTROL YOUR BODY

FOUR TO SIX WEEKS
Thyroxine can stay in the body for over a month

50
The number of different types of hormones in the body

100 YEARS
The term 'hormone' was coined a century ago

The brain secretes atrial and natriuretic peptides, which decrease blood pressure

500 TO 900MG
The pituitary gland can weigh just shy of a gram

All mammals, birds and fish have an endocrine system

Swollen thyroid glands act as a warning of a hormonal imbalance and potential disease

Did you know? Diabetes was once diagnosed by tasting urine

Thyroidism

This can present itself as hyperthyroidism or hypothyroidism. Both conditions concern the thyroid gland and the production of the thyroid hormone, but they are very different diseases. Hyperthyroidism occurs when the thyroid gland secretes too much thyroid hormone.

The thyroid becomes overactive, and this causes the person to be overly nervous with an increased heart rate. They tend to lose weight without trying to and will often sweat excessively. This can sometimes be caused by an autoimmune disorder known as Graves' disease. Hypothyroidism, on the other hand, occurs when the thyroid gland is underactive. The body becomes sluggish and the gland produces too little of the thyroid hormone. The patient becomes extremely tired and other bodily functions slow down. For example, they can suffer from intestinal issues such as severe constipation. The skin can become very dry and patients will often become depressed. It will also result in weight gain.

Spontaneous Cushing's Syndrome

Cushing's syndrome occurs when your body makes too much cortisol hormone, which is made by the adrenal glands. This can happen if the pituitary gland is aggravated by something, such as a benign tumour. The pituitary gland releases too much adrenocorticotropic hormone (ACTH), which is sent to the adrenal glands via the bloodstream. The ACTH commands the adrenal glands to make extra cortisol that the body doesn't need. Symptoms include weak bones, sudden weight gain and a darkening of the skin's pigmentation. The problem can usually be solved by removing the tumour.

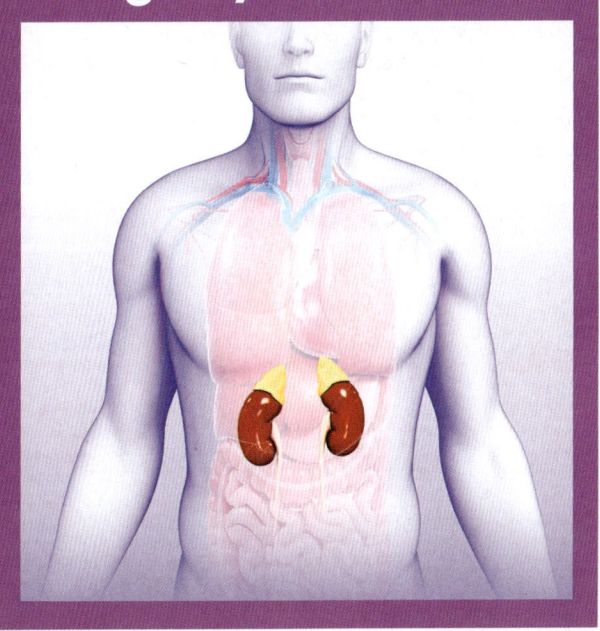

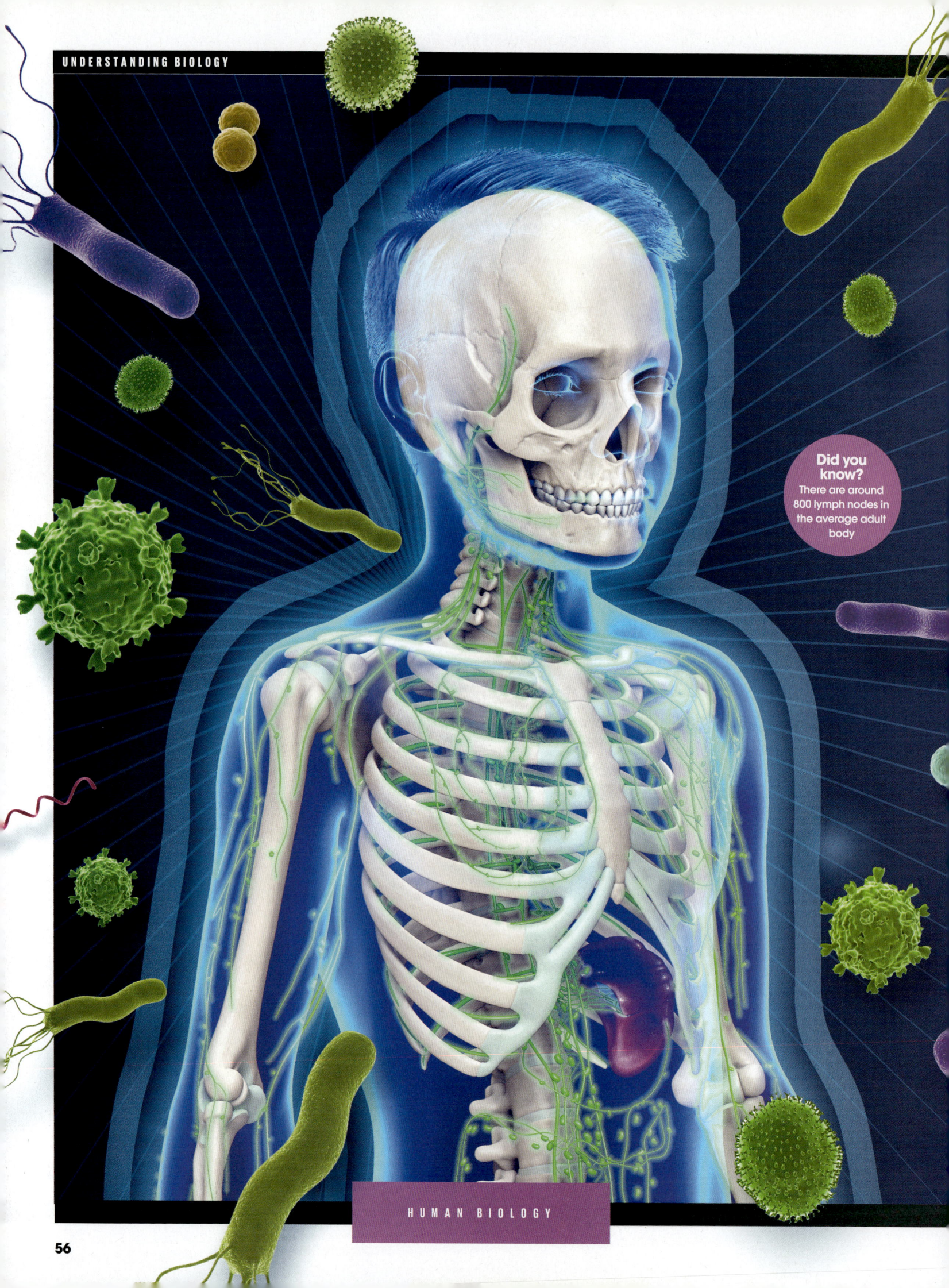

UNDERSTANDING BIOLOGY

Did you know? There are around 800 lymph nodes in the average adult body

HUMAN BIOLOGY

YOUR IMMUNE SYSTEM EXPLAINED

Your immune system explained

Meet the cells and organs that make up your immune system. Discover how it keeps deadly invaders at bay and what happens when it gets out of control

Every day your immune system is working tirelessly to fend off harmful invading pathogens such as bacteria and viruses. Our understanding of the immune system began in the late-19th century when Russian zoologist Élie Metchnikoff identified a group of white blood cells called phagocytes, whose purpose is to seek out, engulf and eliminate pathogens. During the same period, the discovery of antibodies and their role in neutralising pathogens also came to light thanks to German physiologists Emil Behring and Paul Ehrlich. Since then, scientists from around the world have come to understand the complexity of our immune system and the many ways it has evolved to fight off bacterial baddies and villainous viruses.

Part of our immune system is passed on to us as newborns from our mothers through the placenta during our time spent in the womb, and again through feeding on breast milk. Immunity to particular viruses and bacteria and the potency of this passive immunity differs from mother to mother. For example, mothers who have had chickenpox during their lifetime can pass on that immunity to their newborn. However, this form of passive protection is short-lived and begins to decrease after the first few months of life. Over time, the immune system acquires more complex immunities after reacting with pathogens from the outside world.

UNDERSTANDING BIOLOGY

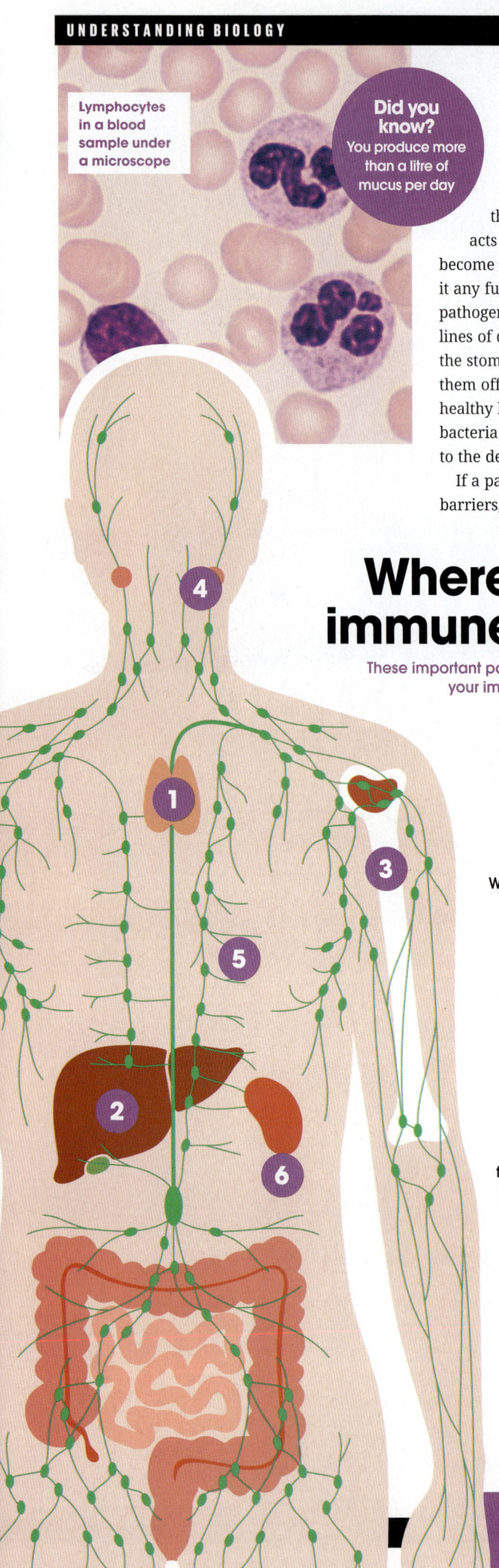

Lymphocytes in a blood sample under a microscope

Did you know? You produce more than a litre of mucus per day

Your immune response is divided into systems. The first frontline defence is the innate immune system. The skin is our first defence against pathogens and offers a physical barrier around the body. Then there's mucus. Snot acts like quicksand for pathogens, which become stuck and are prevented from making it any further into the body. For those sneaky pathogens that make it past these first two lines of defence, they might find their way into the stomach, which is filled with acid that kills them off. Also, the digestive system is home to healthy bacteria that compete with invading bacteria for space and food, sometimes leading to the demise of the invading bacteria.

If a pathogen makes it past these physical barriers, then a secondary internal defence is called upon. In the event that a physical barrier, such as the skin, has been broken and there's a potential threat of invasion from outside pathogens, white blood cells called mast cells, which are found in the connective tissue near blood vessels, release histamine into the blood to initiate the body's inflammatory response.

The first soldiers to battle pathogens are a group of white blood cells called phagocytes. The most abundant phagocytes in the human body are called neutrophils, which gobble up pathogens on sight. Once a pathogen is devoured, the neutrophil essentially self-destructs in a process called apoptosis, killing the pathogen in the process. During infection, the bodies of self-destructed neutrophils pile up and form pus at the site of a wound.

Where is your immune system?

These important parts of the body make up your immune system

1 Thymus
White blood cells called lymphocytes arrive at the thymus from the bone marrow and mature into T cells.

2 Liver
White blood cells called phagocytes are stored here. The liver can also detect when pathogens enter the body.

3 Bone marrow
This is the factory for all immune cells, which begin as immature stem cells and diversify into specialised immunity cells.

4 Tonsils
A collection of lymphocytes can be found in the tonsils. These cells are one primary defence against pathogens.

5 Lymph Nodes
Immunity cells called B cells and T cells gather in the lymph nodes to communicate with each other.

6 Spleen
This filters out pathogens such as bacteria and viruses from the blood, and also detects faulty blood cells.

Herd immunity

During the height of the coronavirus pandemic, you might have heard the term 'herd immunity' being used. This form of immunity occurs when the majority of a population, or 'the herd', is immune to a particular infectious disease. At the point at which the herd is predominantly immune, the disease is less likely to spread. For example, if a person infected with highly contagious measles were to stand in the centre of a group of vaccinated people, the disease would be unable to be transmitted and would eventually disappear. This type of immunity also helps those that are unable to be vaccinated, such as newborns and the elderly.

For herd immunity to be effective, the people vaccinated or immune to the disease need to outnumber the rate at which the disease can spread. Measles, for example, can spread at such a rapid rate that 19 out of 20 people need to be vaccinated for herd immunity to be achieved.

For herd immunity to work, the majority of the population needs to be vaccinated

HUMAN BIOLOGY

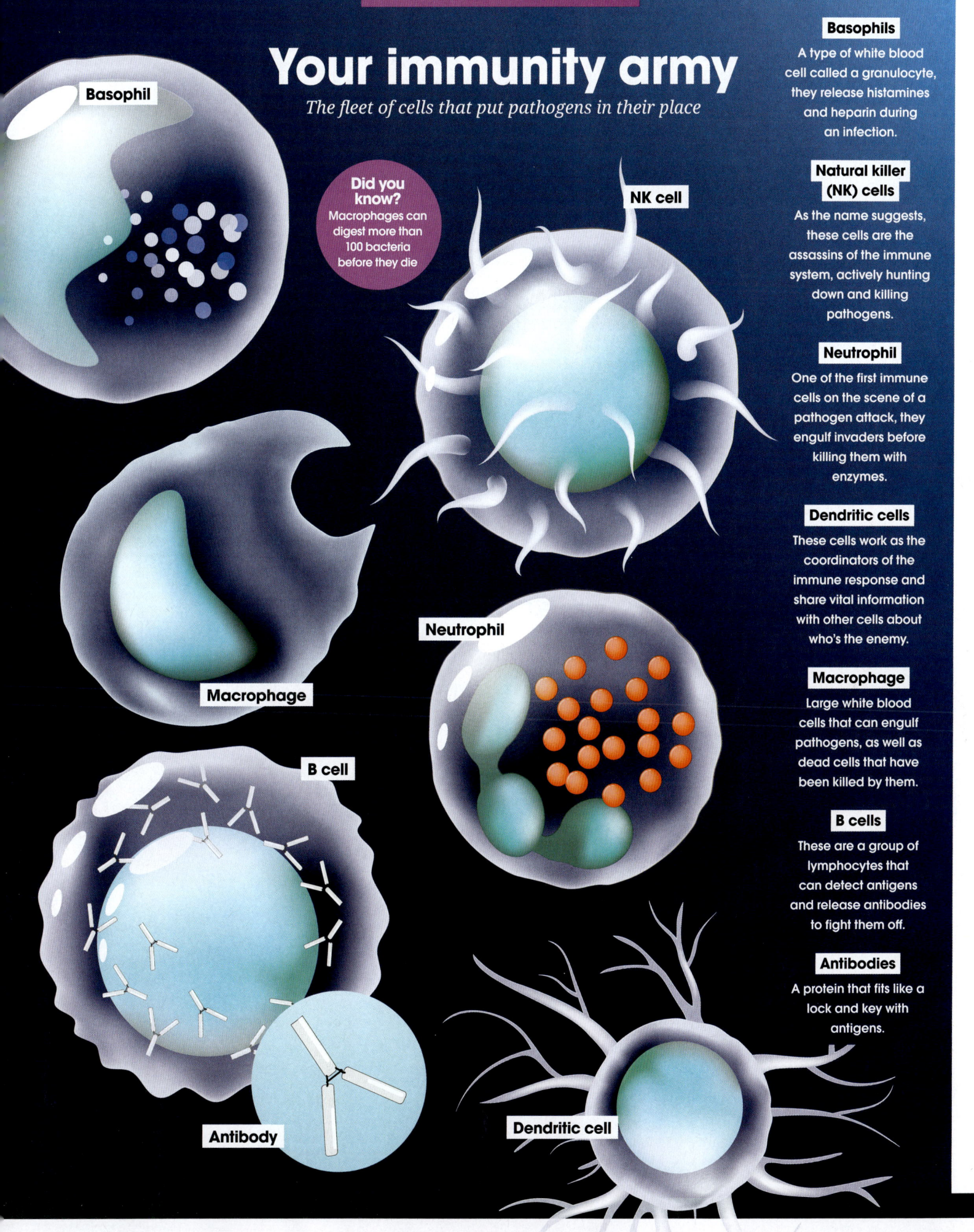

YOUR IMMUNE SYSTEM EXPLAINED

Your immunity army
The fleet of cells that put pathogens in their place

Did you know? Macrophages can digest more than 100 bacteria before they die

Basophil

NK cell

Macrophage

Neutrophil

B cell

Antibody

Dendritic cell

Basophils
A type of white blood cell called a granulocyte, they release histamines and heparin during an infection.

Natural killer (NK) cells
As the name suggests, these cells are the assassins of the immune system, actively hunting down and killing pathogens.

Neutrophil
One of the first immune cells on the scene of a pathogen attack, they engulf invaders before killing them with enzymes.

Dendritic cells
These cells work as the coordinators of the immune response and share vital information with other cells about who's the enemy.

Macrophage
Large white blood cells that can engulf pathogens, as well as dead cells that have been killed by them.

B cells
These are a group of lymphocytes that can detect antigens and release antibodies to fight them off.

Antibodies
A protein that fits like a lock and key with antigens.

UNDERSTANDING BIOLOGY

Fighting infections
How your immune system actively responds to a pathogen invasion

Pathogen

10 Assassins
T helper cells detect the MHC presented on the phagocyte's surface and release signalling proteins called cytokines to activate NK cells to destroy infected cells.

1 B cell
These are covered with antibodies. When they come into contact with an antigen in the blood or the spleen, they immediately bind to it.

9 Show and tell
When a phagocyte engulfs a pathogen, it presents parts of the pathogen on its surface, called a major histocompatibility complex (MHC).

Did you know?
Neutrophils make up around 70 per cent of your white blood cells

4 Memory cells
Some B cells stay away from the fighting to preserve their antibodies in case of a future attack.

7 Call in the big guns
Antibodies have more than one binding and can partly bind with multiple pathogens, acting like glue to clump them together in a process known as agglutination.

11 Infect the infected
Once an NK cell has spotted an infected cell, it binds with it and releases enzymes into the cell to kill it, along with the pathogen within.

8 Mealtime
Macrophages are capable of recognising and completely engulfing either clumped-up pathogens or dead infected cells through a process called phagocytosis.

HUMAN BIOLOGY

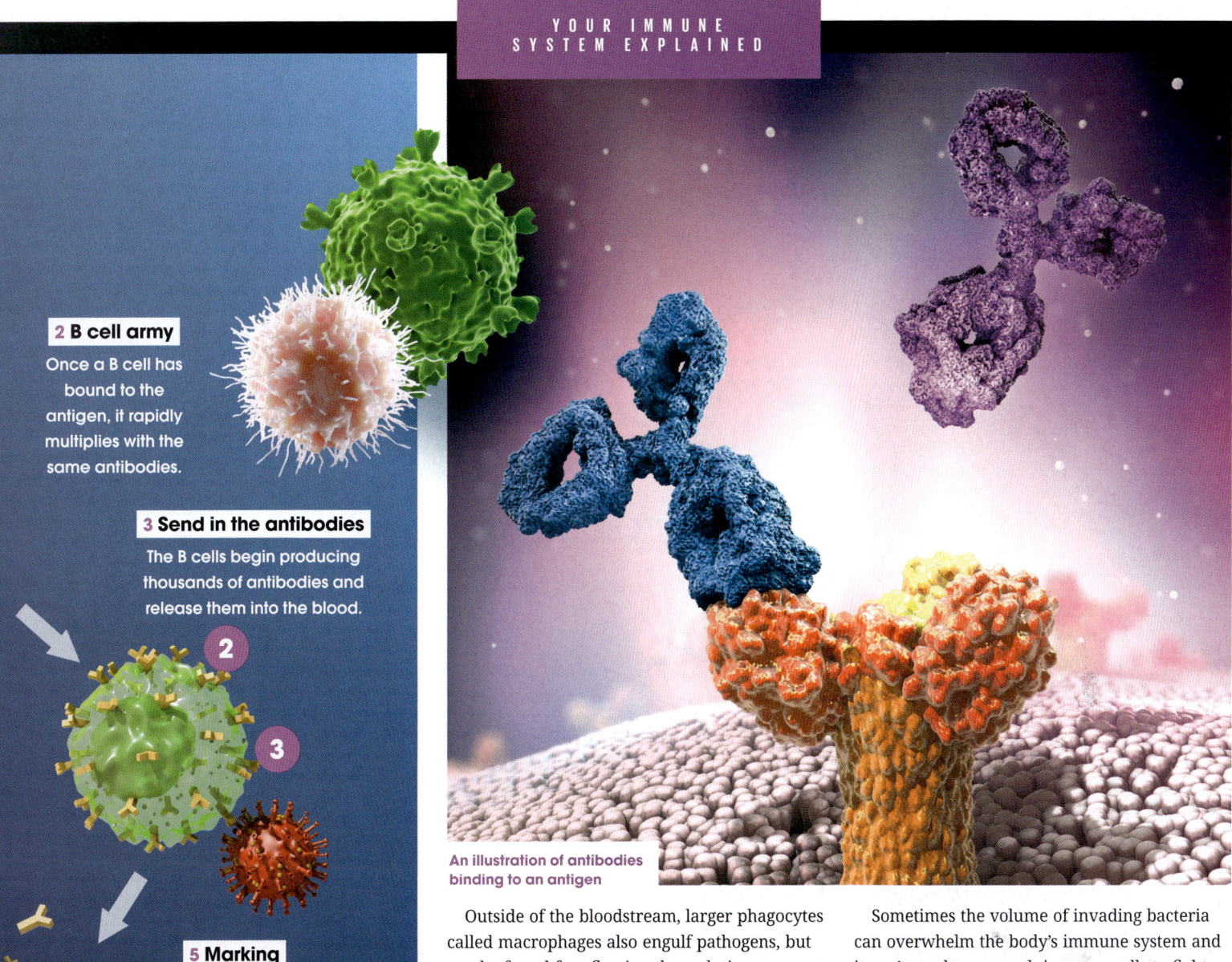

YOUR IMMUNE SYSTEM EXPLAINED

2 B cell army
Once a B cell has bound to the antigen, it rapidly multiplies with the same antibodies.

3 Send in the antibodies
The B cells begin producing thousands of antibodies and release them into the blood.

5 Marking the intruder
The antibodies seek out the matching pathogen and bind to it.

6 Neutralisation
By completely covering the pathogen with antibodies, the pathogen is unable to bind to healthy cells and infect them.

An illustration of antibodies binding to an antigen

Outside of the bloodstream, larger phagocytes called macrophages also engulf pathogens, but can be found free-flowing through tissue or embedded into tissues such as the lymph nodes or intestinal tract. If the war against pathogens isn't going in the phagocytes' favour, they can release chemicals into the blood called pyrogens to trigger a fever. Once the pyrogens reach the brain's hypothalamus, they cause it to raise the body's temperature, giving your immune cells a metabolic boost for battle but inducing stress in pathogens.

In the event that these innate defences can't completely fend off pathogens, the second division of the immune system, known as adaptive or active immunity, kicks into gear. In this system, a legion of white blood cells called lymphocytes work together to seek out and destroy pathogens that enter the bloodstream. The key elements in the active immune system are antigens and antibodies. Antigens are like the identification cards for pathogens and antibodies are like pathogen mugshots. When the two are compared and a positive match is found, immune cells are alerted and the elimination process begins.

Sometimes the volume of invading bacteria can overwhelm the body's immune system and it can't produce enough immune cells to fight off a serious infection, such as pneumonia. In this case, the immune system may need a helping hand from medication, called antibiotics, to eradicate bacterial pathogens. Antiviral medication also works to limit the infectious abilities of some invading viruses, giving the immune system a chance to defeat them.

Similarly, vaccines are used to stimulate our natural immune response against specific pathogen threats. Vaccines work by releasing proteins called antigens into the body, which the immune system recognises as foreign and a potential pathogen. The antigen in the vaccine will have the same proteins as that of the pathogen you're vaccinating against, such as measles. This sets the active immune system into motion, creating antibodies to fight a potential infection. The immune system will store or 'remember' the antibodies it needs in case of a future infection. Many vaccines are administered yearly to keep the system primed for defence.

Did you know?
70 per cent of immunity happens in the digestive tract

61

UNDERSTANDING BIOLOGY

When the immune system turns on itself

Sometimes the human immune system misinterprets healthy tissue as invading pathogens and sets out to remove it. In a typical immune response, T cells release cytokines that instruct other immune cells when and where to attack and when to fall back. But sometimes this cellular communication goes wrong and the immune cells can't correctly identify what's a pathogen and what's healthy tissue, which leads to the development of autoimmune diseases such as lupus and rheumatoid arthritis.

There are more than 100 known autoimmune diseases, some more serious than others. Multiple sclerosis, for example, is a chronic disease of the central nervous system caused by an autoimmune response. Instead of solely fighting invading pathogens, immune cells attach to nerve cells and prevent them from sending electrical signals to and from the brain. It's still unclear why autoimmune diseases develop and what causes the immune system to stop recognising healthy tissue. However, scientists have made some connections between autoimmune diseases and possible triggers, such as an immune response to an infection like strep throat that can lead to the development of an autoimmune disease like psoriasis.

Injury or damage to a part of the body may also trigger the development of a disease such as psoriatic arthritis, which affects the joints. Along with some environmental factors, genetics may also play a pivotal role in the development of some autoimmune diseases.

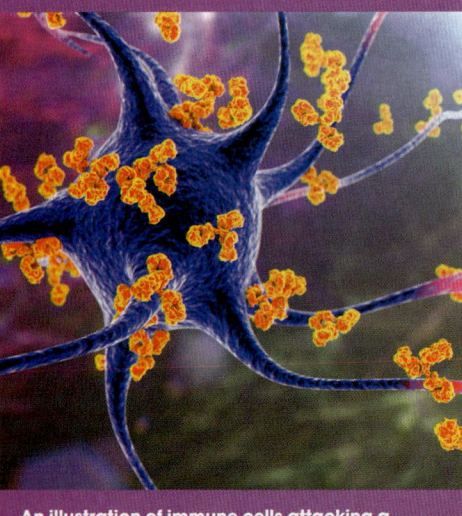

An illustration of immune cells attacking a healthy nerve cell

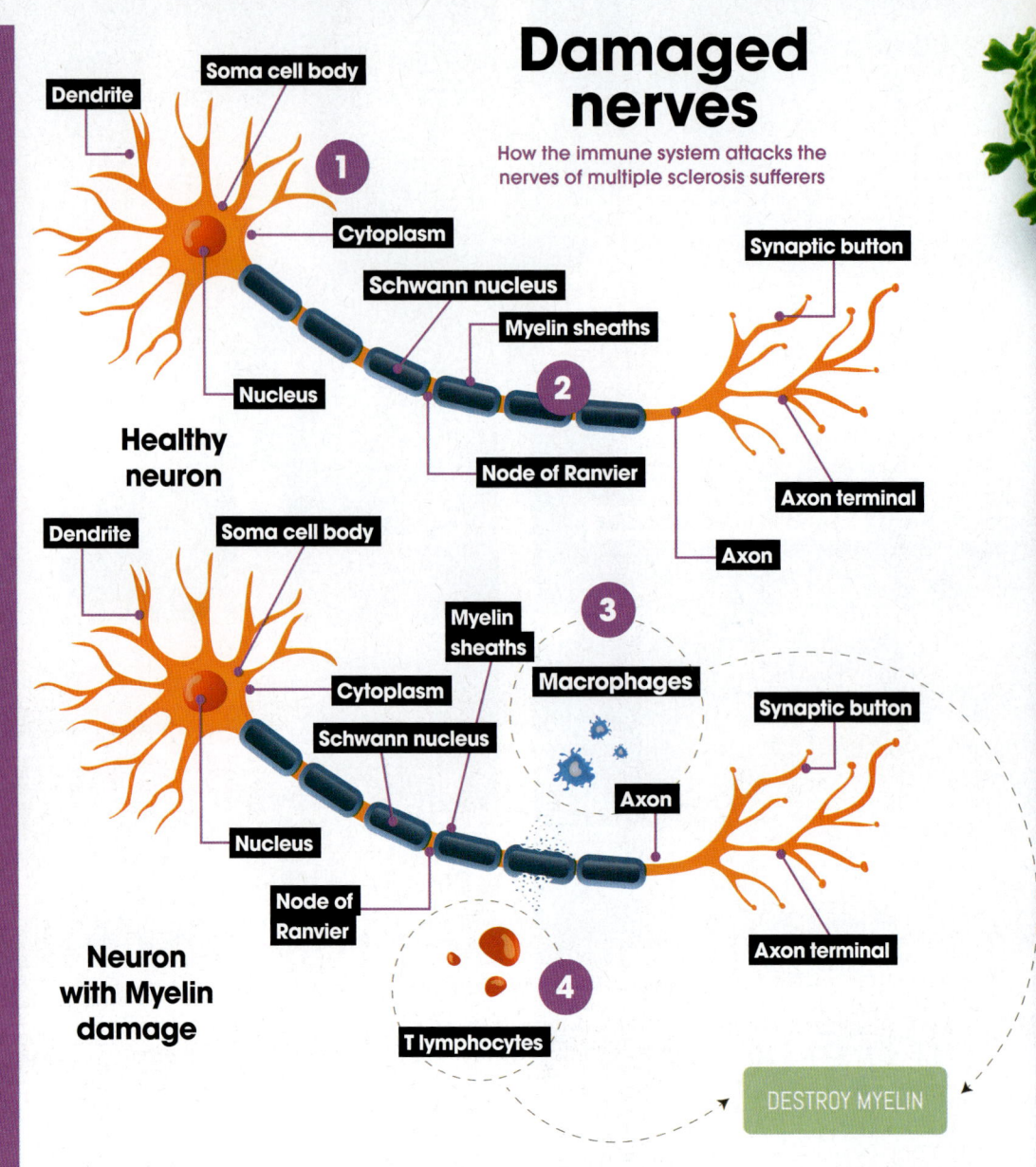

Damaged nerves
How the immune system attacks the nerves of multiple sclerosis sufferers

Healthy neuron — Dendrite, Soma cell body, Cytoplasm, Schwann nucleus, Myelin sheaths, Nucleus, Node of Ranvier, Synaptic button, Axon terminal, Axon

Neuron with Myelin damage — Dendrite, Soma cell body, Cytoplasm, Schwann nucleus, Myelin sheaths, Nucleus, Node of Ranvier, Macrophages, Axon, Synaptic button, Axon terminal, T lymphocytes → DESTROY MYELIN

1 Healthy neuron
These are specialised cells that send information around the nervous system through electrical signals.

2 Myelin
This fatty membrane wraps around the nerve for protection and allows electrical signals to pass along the cell.

3 Destroy message
T cells send the alarm to macrophages to seek and destroy the neuron's myelin.

4 Inflammation
Macrophages and other lymphocytes damage the myelin and create inflammation along the nerve.

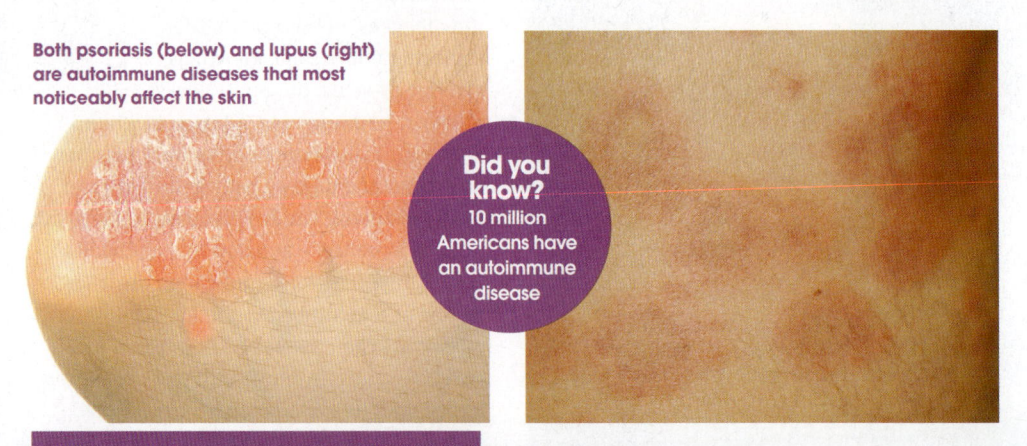

Both psoriasis (below) and lupus (right) are autoimmune diseases that most noticeably affect the skin

Did you know? 10 million Americans have an autoimmune disease

HUMAN BIOLOGY

YOUR IMMUNE SYSTEM EXPLAINED

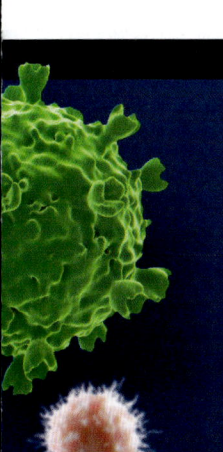

Immunity myths

Sheena Cruickshank, an immunologist and a professor of biomedical sciences and public engagement at the university of Manchester, separates immunity facts from fiction

Does stress lower your immune response?
Chronic stress does, so the longer you're stressed, it absolutely does. I think a short, sharp stress isn't damaging for your immune system. Some people suggest it might give you a little transient boost. But chronic stress, if you stress for a long period of time, does impede the actions of immune cells.

Does immunity weaken with age?
Yes, unfortunately it does. As we get older, the majority of us will see a decline in our immune function. We may have fewer of the specialised white blood cells called lymphocytes that are there to deal with infection. However, it's not always as severe in some older people; some older people do age well. And one thing that might be linked to that is having a good, diverse microbiome. It helps maintain the barrier in your gut, and that stops you getting kind of a leaky gut and getting this low-level inflammation that can happen when you're older, and that can further make your immune system a bit off.

Does vitamin C boost the immune system?
The evidence for that is rather mixed. A healthy diet is important for an optimal immune system. But studies on vitamin C have been really mixed. There's not a really clear link with vitamin C per se. The exception is vitamin D. A lot of people can be vitamin D deficient, particularly in the winter months when we get less sunlight. And you have to eat quite a lot of particular food types to really try and enrich that in your diet. And people who have low vitamin D do seem to be a bit more susceptible to catching colds.

Is the idea of being able to boost your immune system a myth?
There is one way that you can truly boost your immune system, and that's having a vaccine. I mean, that's what a vaccine is doing – it's boosting your immunity to stop you getting sick in response to a particular infection. But the idea that we can boost our general immunity is a myth. For most of us, our immune system is doing a pretty good job. I mean, we're exposed to infections and nasty germs every day, and most of the time we don't get ill.

Does sleep have any impact on our immune system?
It's important to have sleep for our immune systems, like all sorts of cells in our body. Pretty much all the cells in the body have little circadian clocks and they help kind of optimise our function, so lack of sleep can throw those out, and it can definitely affect your immune response. And we know it ourselves. If we've been going through a phase or we've not been sleeping well and we've been stressed or we've not been eating well, then we do often get run down; we'll often get that horrible infection taking hold.

Does exercise lower your immune system?
Now, it all depends how you exercise. Moderate exercise is a benefit for your immune system, so moderate levels of activity, that's kind of what the NHS is recommending. Three or four bursts of moderate kinds of exercise where you can really feel your heart starting to get going. The 10,000 steps a day was thought to be a bit of a kind of random figure, but actually it's been assessed since, and that does look like a good thing to aim for. So all of these things seem to be good at mobilising your immune system, helping it operate better, detect infection, deal with infections better, and you seem to have a lower risk of upper respiratory tract infections, like colds. But if you do extreme sports and you're absolutely not used to it, that puts your body under stress and that is not so good for your immune system. It's that balance, and also how acclimatised you are too.

What's a common myth you often debunk?
One of the things that really worries me that I'm seeing a lot more of at the moment is this idea that it's good for children to get infected and build their immune system. I don't really understand where that narrative has come from. Actually, we know that children under the age of five are some of the most vulnerable to infection.

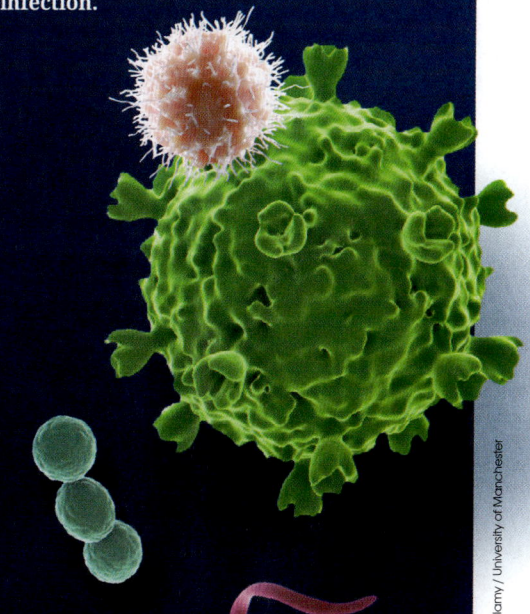

Did you know?
The thymus shrinks at a rate of three per cent per year until middle age

63

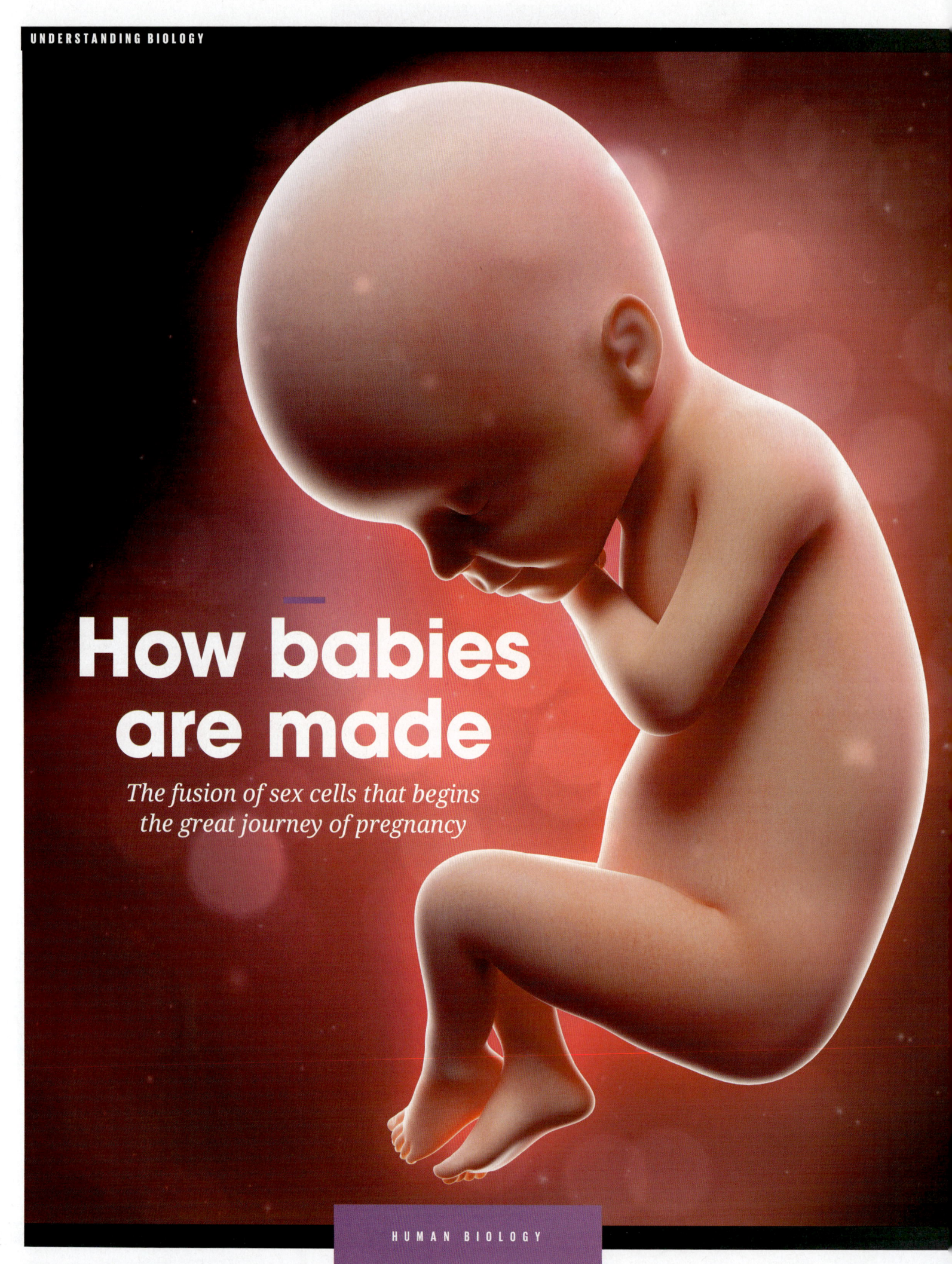

UNDERSTANDING BIOLOGY

How babies are made

The fusion of sex cells that begins the great journey of pregnancy

HUMAN BIOLOGY

HOW BABIES ARE MADE

Humans reproduce by sexual reproduction. This mixes the genetic information of two people. For this to happen we have evolved specialised sex cells, known as gametes, that are specific to each sex. Females produce and carry egg cells that, much like their vastly larger avian equivalents, contain both genetic information and form the nexus of an embryo. Females also have a dedicated sex organ called a uterus, which nurtures the embryo during its development. Males provide the other half of the genetic information by producing sperm cells.

Both the egg and sperm hold instructions encoded by deoxyribonucleic acid (DNA), which contains all the information needed to form a functional human body. The cells' DNA is wrapped up tightly into large structures called chromosomes. Both the egg cell and the sperm cell carry 23 chromosomes, which once aligned in an embryo will form 23 pairs that will be carried by every somatic cell in a mature human body. Two of these chromosomes form a pair known as sex chromosomes, as they define the sex of the baby during development. Every egg cell carries an X chromosome, and a sperm cell can carry either an X or a Y chromosome. Females are encoded by a combination of XX, and males by a combination of XY. Our genetic sex is sealed the very moment our two sets of chromosomes meet during fertilisation.

Fertilisation can begin once an egg is released by a follicle in the ovaries, a step that occurs periodically in women of fertile age. Throughout this cycle, stages of hormones trigger the release of an egg, which in turn triggers more signals intended for the wall of the uterus, which thickens in expectation of a fertilised embryo to implant in its surface. This offers a brief window of time for legions of sperm to be released from a male's testicles and undertake the long, arduous journey through a female's uterus to reach the egg while it is viable.

The fusion between a sperm and egg cell's DNA marks the beginning of embryo development, and once it has found its home in the nourishing lining of the uterine wall, the embryo undergoes a monumental growth spurt. It transforms from a mere bundle of cells into a foetus, acquiring a heart, brain, jaw, fingers and toes. Within the first 12 weeks of pregnancy the foetus boasts a full complement of organs. This incredible metamorphosis is guided on a cellular level by genetic instructions encoded within the DNA and by external environmental signals. These signals tell the cells how to interact with one another, where to migrate, how to divide and when to die to make way for new cells. This allows a swarm of microscopic cells to differentiate and develop into something not only gargantuan in size but also in complexity: a fully formed human baby.

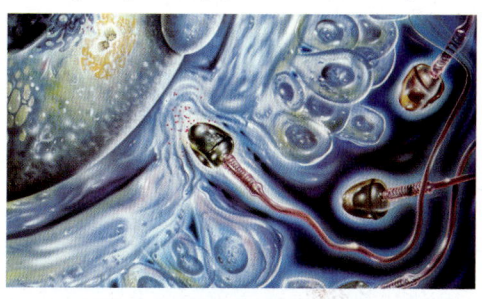

A sperm cell must pass a thick membrane encasing the egg, known as a zona pellucida, before fertilisation can occur

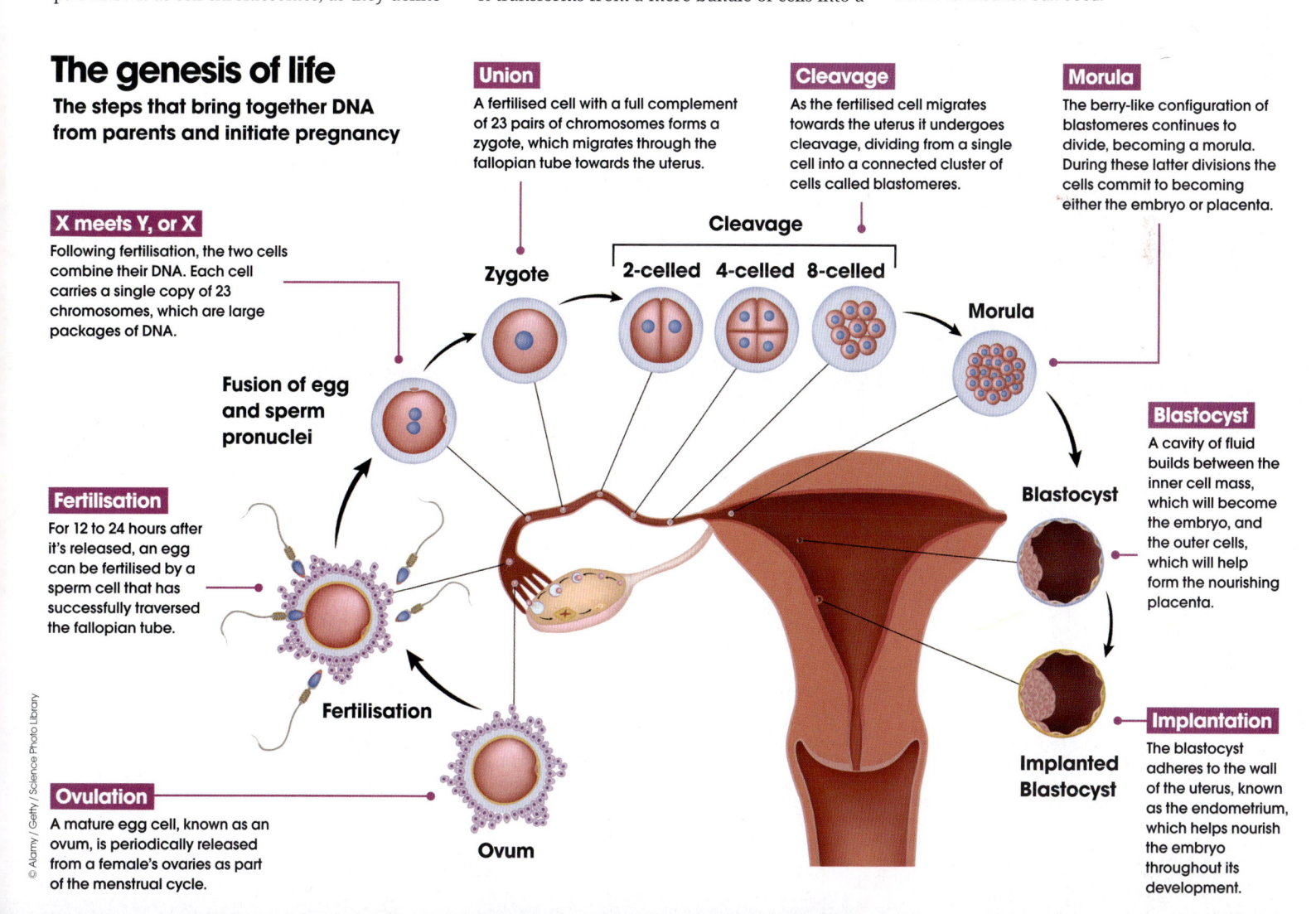

The genesis of life
The steps that bring together DNA from parents and initiate pregnancy

X meets Y, or X
Following fertilisation, the two cells combine their DNA. Each cell carries a single copy of 23 chromosomes, which are large packages of DNA.

Fertilisation
For 12 to 24 hours after it's released, an egg can be fertilised by a sperm cell that has successfully traversed the fallopian tube.

Ovulation
A mature egg cell, known as an ovum, is periodically released from a female's ovaries as part of the menstrual cycle.

Union
A fertilised cell with a full complement of 23 pairs of chromosomes forms a zygote, which migrates through the fallopian tube towards the uterus.

Cleavage
As the fertilised cell migrates towards the uterus it undergoes cleavage, dividing from a single cell into a connected cluster of cells called blastomeres.

Morula
The berry-like configuration of blastomeres continues to divide, becoming a morula. During these latter divisions the cells commit to becoming either the embryo or placenta.

Blastocyst
A cavity of fluid builds between the inner cell mass, which will become the embryo, and the outer cells, which will help form the nourishing placenta.

Implantation
The blastocyst adheres to the wall of the uterus, known as the endometrium, which helps nourish the embryo throughout its development.

UNDERSTANDING BIOLOGY

In Vitro Fertilisation
How embryos can be created in a lab

IVF allows scientists to microscopically examine and genetically screen embryos for abnormalities before implantation

The process of fertilisation and embryo implantation is a wonderfully elegant yet complex process. This complexity, however, presents numerous opportunities for natural obstacles and barriers that prevent natural conception. Such obstacles include damage or blockage of the fallopian tubes, which can prevent the sperm and egg meeting or the fertilised egg from reaching the uterus. Other disorders affect the release of eggs from the ovaries, and others affect males by reducing sperm number and activity. These challenges have driven the development of in vitro fertilisation (IVF), which offers a human-made bridge to pregnancy when the natural path is blocked. IVF is a process whereby the act of fertilisation, initial embryo division and implantation are conducted in the laboratory under the guidance of scientists.

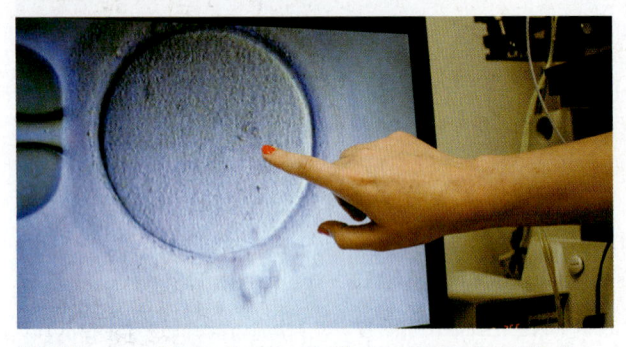

IVF allows scientists to microscopically examine and genetically screen embryos for abnormalities before implantation

Sperm injection

Males with semen that contains millions of healthy and active sperm cells per millilitre are often able to achieve fertilisation organically. But semen harbouring a low sperm count, irregular-shaped sperm or sperm that aren't particularly mobile can find it difficult, or even impossible. Sperm injection provides an alternative means to achieve fertilisation. Clinicians can use micropipettes to hold the egg in place and directly inject sperm cells into the egg using a micro-sized needle, allowing the sperm cell to reach its goal.

Tiny needles can be used to penetrate the egg and directly inject sperm

The world's first 'test tube' baby

In 1977, Lesley Brown – who was struggling with infertility due to blocked fallopian tubes – was put in contact with scientists Dr Robert Edwards and Dr Patrick Steptoe. Steptoe was an expert in obtaining eggs from ovaries and Edwards an expert in fertilising human eggs in a petri dish. Together the two offered Lesley Brown the opportunity – albeit one with a slim chance of success – to artificially fertilise her eggs with her husband's sperm. Without the use of hormones to manipulate her natural menstrual cycle and increase egg production, the scientists overcame the odds and successfully performed the first in vitro fertilisation and implantation of a single lab-fertilised embryo.

Fertilisation in a laboratory
The artificial fertilisation technique that helps parents overcome natural barriers to conception

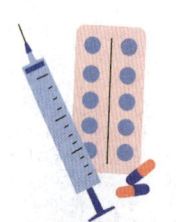

1. Increasing egg supply
Hormones are administered which suppress the natural menstrual cycle. These are followed by fertility hormones that boost egg production by the ovaries.

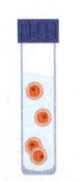

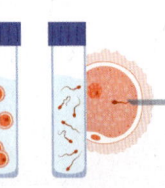

2. Collection and fertilisation
Eggs are collected using a needle inserted into each ovary and fertilised with sperm by either mixing or directly injecting the sperm into the egg.

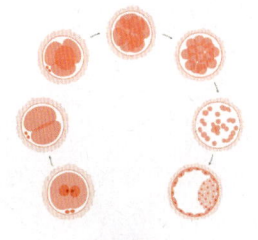

3. Embryo development
The embryo is stored in an incubator and begins to divide. Many clinics wait until the cell has divided into a blastocyst before implantation.

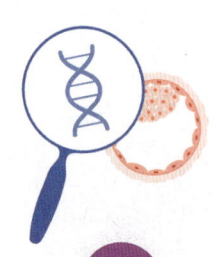

4. Genetic screening
DNA from a single or small number of cells is removed from the embryo and tested for abnormalities.

5. Cryopreservation
Healthy embryos that will not be implanted immediately can be frozen and stored safely for years for later implantation.

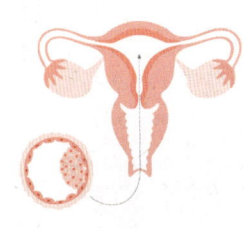

6. Embryo transfer
The embryo is implanted in the uterus using a flexible tube called a catheter, which is carefully guided into place using an ultrasound scan.

HUMAN BIOLOGY

HOW BABIES ARE MADE

One baby, three parents

How doctors can use different genetic material

Nuclear DNA resides inside the nucleus of our cells. This DNA is huge, composed of reams of instructions 3.3 billion base pairs in length. Nuclear DNA comes from both our mother and father and is responsible for the vast majority of characteristics we display as we develop. However, our cells also host a small chunk of independent DNA that lives inside cellular components called mitochondria, which are energy-producing factories that power our cells. Mitochondrial DNA comes solely from our mother and is inherited from the fertilised egg cell at the beginning of gestation.

Mitochondrial DNA is miniscule in length compared to its nuclear counterpart, measuring around 16,500 base pairs in length, and contains instructions only for the function of mitochondria. In rare cases a mother can harbour a catastrophic mutation in this DNA that prevents healthy embryos developing. To circumvent this problem, scientists have developed a technique that combines nuclear DNA from two genetic parents with mitochondrial DNA from a donor. This removes the risk of inheriting mitochondrial diseases and means that the birthed baby will forever carry DNA from three people.

Room for one more

Replacing a mother's diseased mitochondrial DNA allows a baby to carry genes from three people

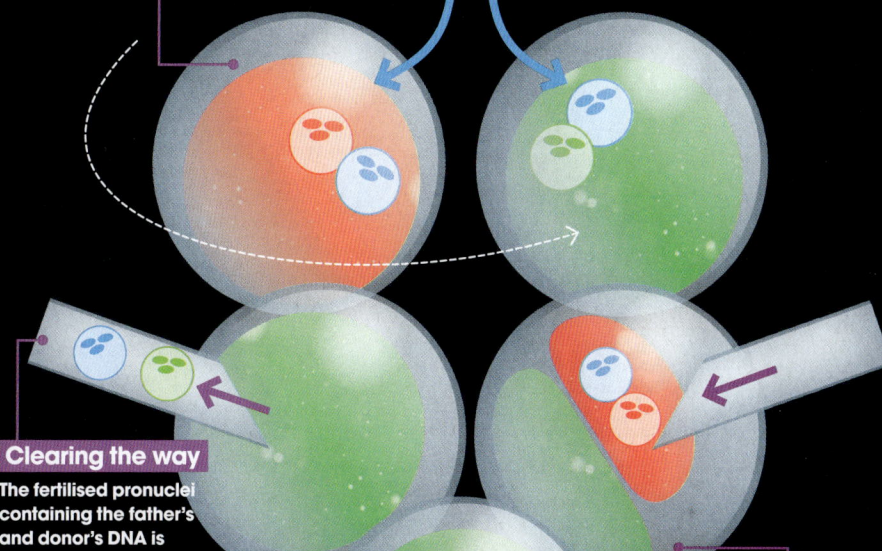

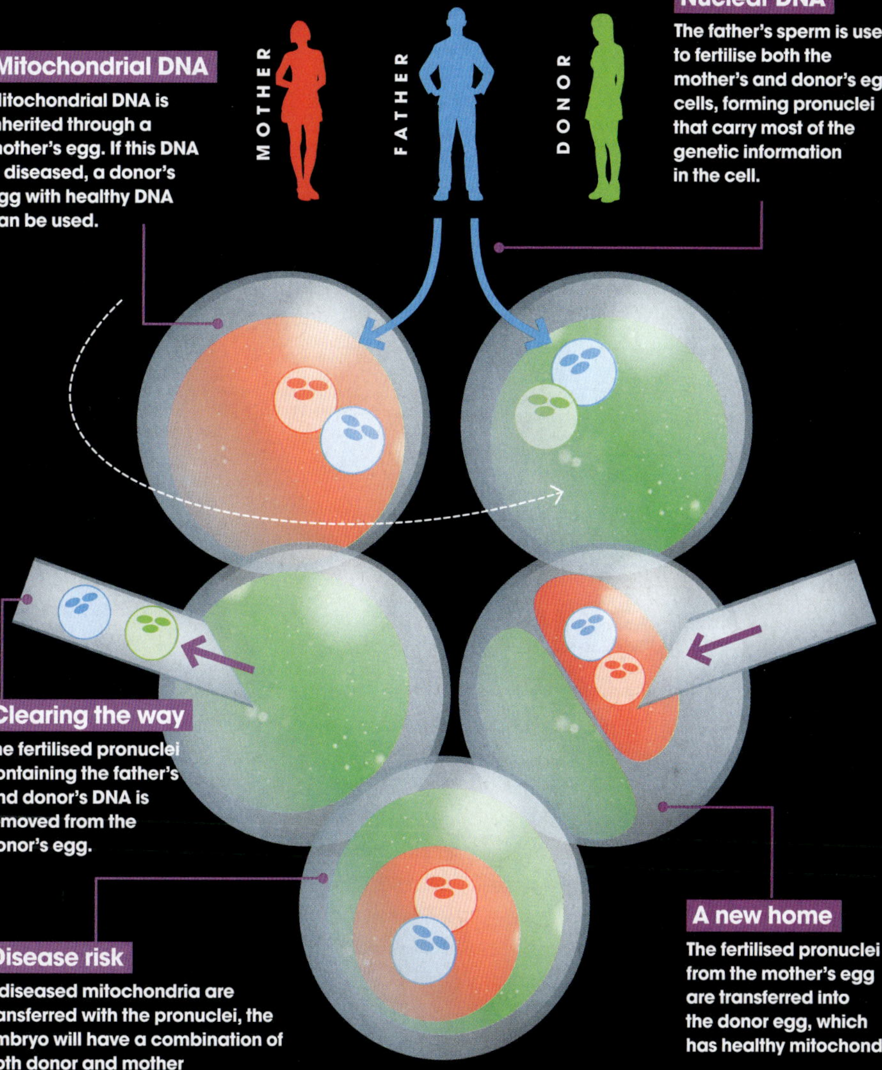

Mitochondrial DNA
Mitochondrial DNA is inherited through a mother's egg. If this DNA is diseased, a donor's egg with healthy DNA can be used.

Nuclear DNA
The father's sperm is used to fertilise both the mother's and donor's egg cells, forming pronuclei that carry most of the genetic information in the cell.

Clearing the way
The fertilised pronuclei containing the father's and donor's DNA is removed from the donor's egg.

Disease risk
If diseased mitochondria are transferred with the pronuclei, the embryo will have a combination of both donor and mother itochondrial DNA.

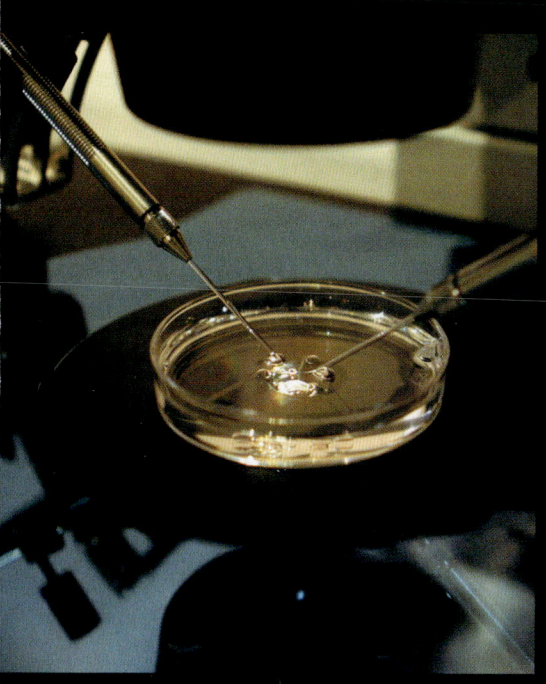

A new home
The fertilised pronuclei from the mother's egg are transferred into the donor egg, which has healthy mitochondria.

Three-parent babies have nuclear DNA from their mother and father and mitochondrial DNA from a donor

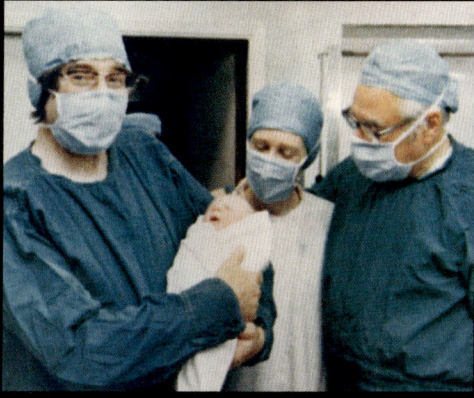

Physiologist Robert Edwards (left) and gynaecologist Patrick Steptoe (right) were the forerunners of successful IVF treatment

Embryos fertilised in the laboratory can be safely cryogenically preserved for over a decade

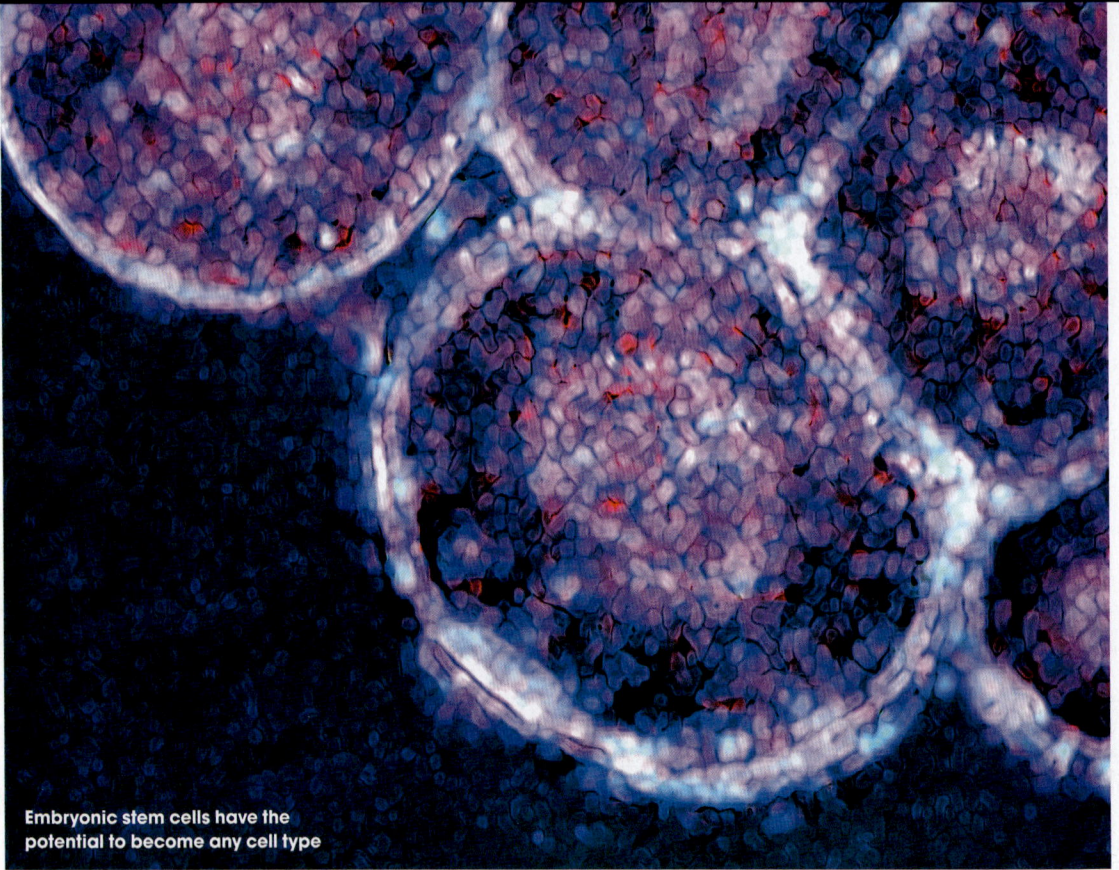

UNDERSTANDING BIOLOGY

Embryonic stem cells have the potential to become any cell type

The science of stem cells

Cloning

The egg cell is essential, designed to ensure that the encoded DNA in the nucleus develops into an embryo. But scientists wondered if the egg could only drive the embryonic development of DNA from both parents, or if it could transform any nuclear DNA into an embryo. In the 1960s they started to find out. Beginning with frog cells, as they're large and easier to manipulate, scientists removed the nuclear DNA from a fertilised egg and replaced it with nuclear DNA from an intestinal cell of an adult frog. After around 40 days, the result was a tadpole that was genetically identical to the frog that had donated the intestinal cell, as all the nuclear DNA had come from one animal.

Dolly the sheep was the first mammal successfully cloned

A fertilised embryo that nestles into the uterine wall holds mammoth amounts of potential. The cells destined to become a foetus at that moment all look alike and number in the mere hundreds. Yet the organism that will arise from this small enclave will one day boast trillions of cells and be composed of a plethora of specialist cell types. How can the multitude of different cell types that make up our eyes, brain, lungs and skin come from such a small number of similar cells? The answer to this exponential increase in complexity and specialisation comes from stem cells.

Every cell in the human body contains all the genetic information needed to perform any role. But for our bodies to develop and function efficiently, we need skin cells to behave like skin, and for muscles to behave like muscles. For this to happen our cells become specialised, or differentiated, into particular cell types, meaning they only use a part of the genetic information available in their DNA.

However, all cell types begin their existence as stem cells, undifferentiated cells that have the potential to become many different cell types. Embryonic stem cells are there at the origin of our developmental journey. As the embryo grows and develops into a foetus, chemical signals received by the stem cells begin their journey of differentiation, sealing their fate to become certain cells by silencing and unlocking specific parts of their DNA.

The malleable power of stem cells also represents remarkable opportunities for those who can harness them. As well as being able to recover embryonic stem cells from early embryos, recent advances have discovered the cellular signals needed to convert differentiated cells back into their unspecialised states. This provides multiple means for scientists to ttransform cells into any type of their choosing. Research is underway to grow entire transplant organs for a patient from their own cells, and we can even utilise stem cells to generate egg and sperm cells that give rise to new life.

Engineering an egg factory

How an embryo or cell can be transformed into a fertilisable egg

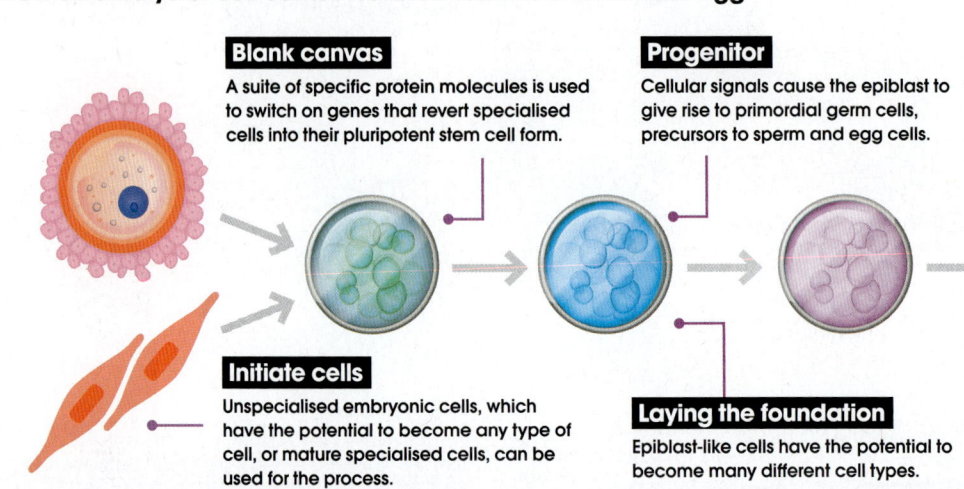

Blank canvas
A suite of specific protein molecules is used to switch on genes that revert specialised cells into their pluripotent stem cell form.

Initiate cells
Unspecialised embryonic cells, which have the potential to become any type of cell, or mature specialised cells, can be used for the process.

Progenitor
Cellular signals cause the epiblast to give rise to primordial germ cells, precursors to sperm and egg cells.

Laying the foundation
Epiblast-like cells have the potential to become many different cell types.

HUMAN BIOLOGY

HOW BABIES ARE MADE

The advantages of cloning

World expert Professor Irina Polejaeva discusses cloning in the agricultural industry

"One key benefit of cloning is that it allows you to introduce specific traits over just one generation. A desired trait may be caused by a specific DNA change and only be naturally present in one breed of cattle and lacking in another. However, we can design clones so that they have attributes from both breeds. For example, horn growth in dairy cattle can be a danger to other animals and those that care for the animals. To induce a lack of horns through a genetic path, you'd usually need to go through multiple generations of breeding, but alternatively we can edit the gene in cells prior to cloning to get this trait immediately. This helps to improve animal welfare as we can avoid farming practices like the dehorning procedure, which is when a calf's horn buds are burned to prevent them growing."

Polejaeva studies the production of transgenic cloned animals for use in agriculture and biomedicine

Supportive scaffolding
The germ cells are aggregated with ovarian tissue, which helps the cells specialise into premature eggs known as oocytes.

In vitro fertilisation
The specialised egg cell is removed from the host and fertilised in the laboratory with sperm cells.

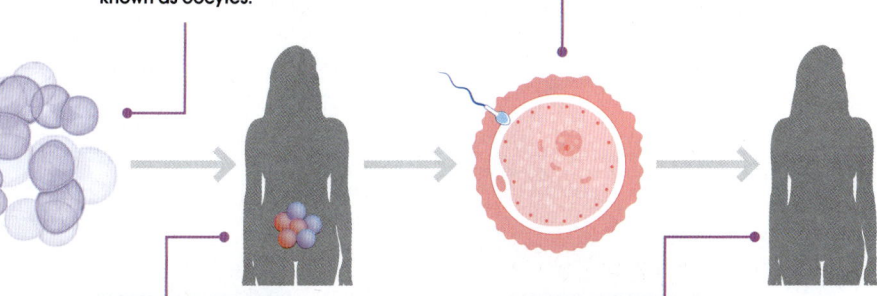

A natural setting
The aggregated cells are implanted into a host's ovary, which enables their continued growth and maturation.

Gestation begins
The once-specialised cell has become part of a fertilised embryo, which is implanted into a donor for gestation.

Embryo growth

What allows embryos to develop such complex structures? We speak with Dr Megan Davey to find out

Davey is a group leader at the Roslin Institute, where she researches chicken embryos to study limb development

The development of an embryo is an amazingly complex phenomenon. Can you outline some of the major proteins involved in driving this process?
One of the main proteins responsible for limb patterns and growth is known as the Sonic hedgehog (SHH) protein. For SHH to work correctly, it has to come on for exactly the right amount of time and have exactly the right amount of activity. If you lose the activity of this protein, embryos develop with unusual differences. Too much can cause too many fingers to form; too little and things like cyclopia, where the eyes don't part and the nose doesn't develop properly, can occur. Another important protein is TALPID3, which affects the activity of centrosomes, cellular structures that are important for cell division. When the cell is not dividing, the centrosome migrates up to the cell surface and docks onto the cell membrane. Cells use centrosomes as their compass, but cells that lack the TALPID3 protein lose their spatial awareness. Centrosomes in these cells continue to migrate, but they don't migrate in the right direction. Instead they move all over. This means that cells can end up growing the wrong way.

Could investigating stem cell activity lead to any medical applications?
We currently don't know how to regenerate fingers, and we don't know if there's a stem cell for regenerating fingers. My hypothesis is that embryos may have stem cells at the tips of their growing digits. We're currently investigating useful genetic tools in chickens, and one that we're developing is called the Brainbow chicken, which allows us to visually mark cells and see where they go during development. We can use this to label the cells found at the end of the digits and see if they behave like stem cells, allowing us to learn more about how to regenerate fingers.

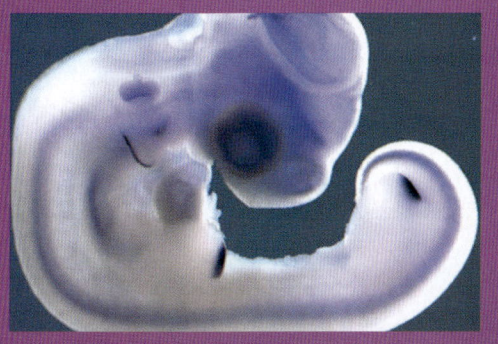

Chicken embryos can be filmed under a microscope to track how embryos develop their limbs

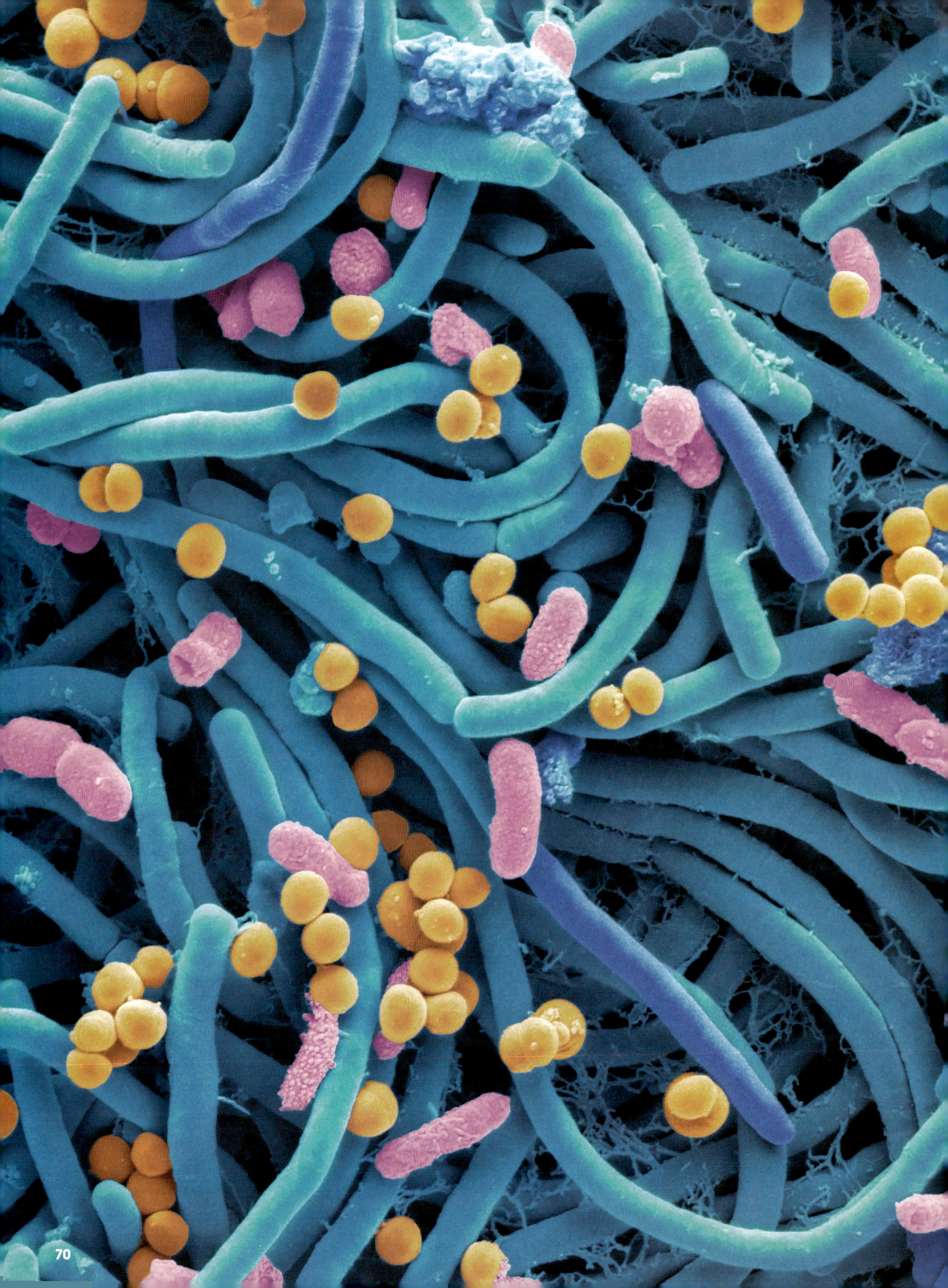

Microbiology

UNDERSTANDING BIOLOGY

The power of pasteurisation

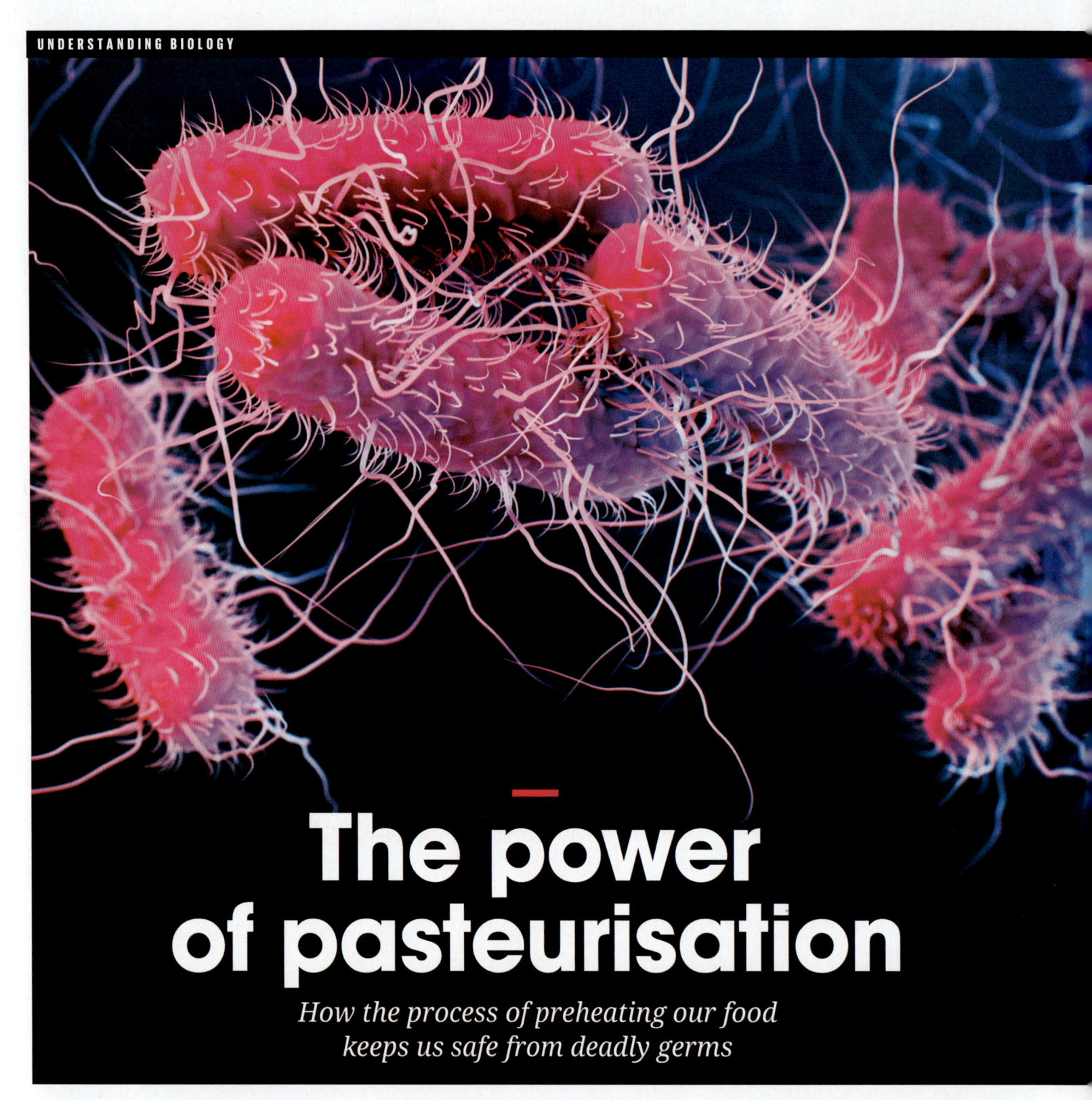

How the process of preheating our food keeps us safe from deadly germs

The process of pasteurisation is a centuries-old method of preparing food, giving it a longer life in larders and today on supermarket shelves. Since the 11th century, people have been using heat to battle against microorganisms to preserve their food. Ancient Chinese wine makers used hot clay jars to heat wine before burying them in cool soil to preserve them. However, it wasn't until the late-19th century that a name was attributed to this process.

Pasteurisation is used to combat the harmful pathogens that reside within raw foods, often referred to as spoilage microorganisms. These can be transferred to our bodies when we consume these foods in their raw state. Pathogens such as E. coli and Salmonella can be found in raw milk, and when ingested can cause serious health issues. To destroy and remove harmful pathogens from certain foods, such as milk, it's heated to high temperatures below 100 degrees Celsius. The heat denatures and breaks down the proteins and enzymes within the pathogens, killing them and allowing humans to consume it without the risk to their health. Once a food has been heated, it's then cooled down and stored. Storing pasteurised products in cold conditions like a fridge creates an environment where microorganisms are unable to reproduce and grow, which would effectively undo the job of pasteurisation.

MICROBIOLOGY

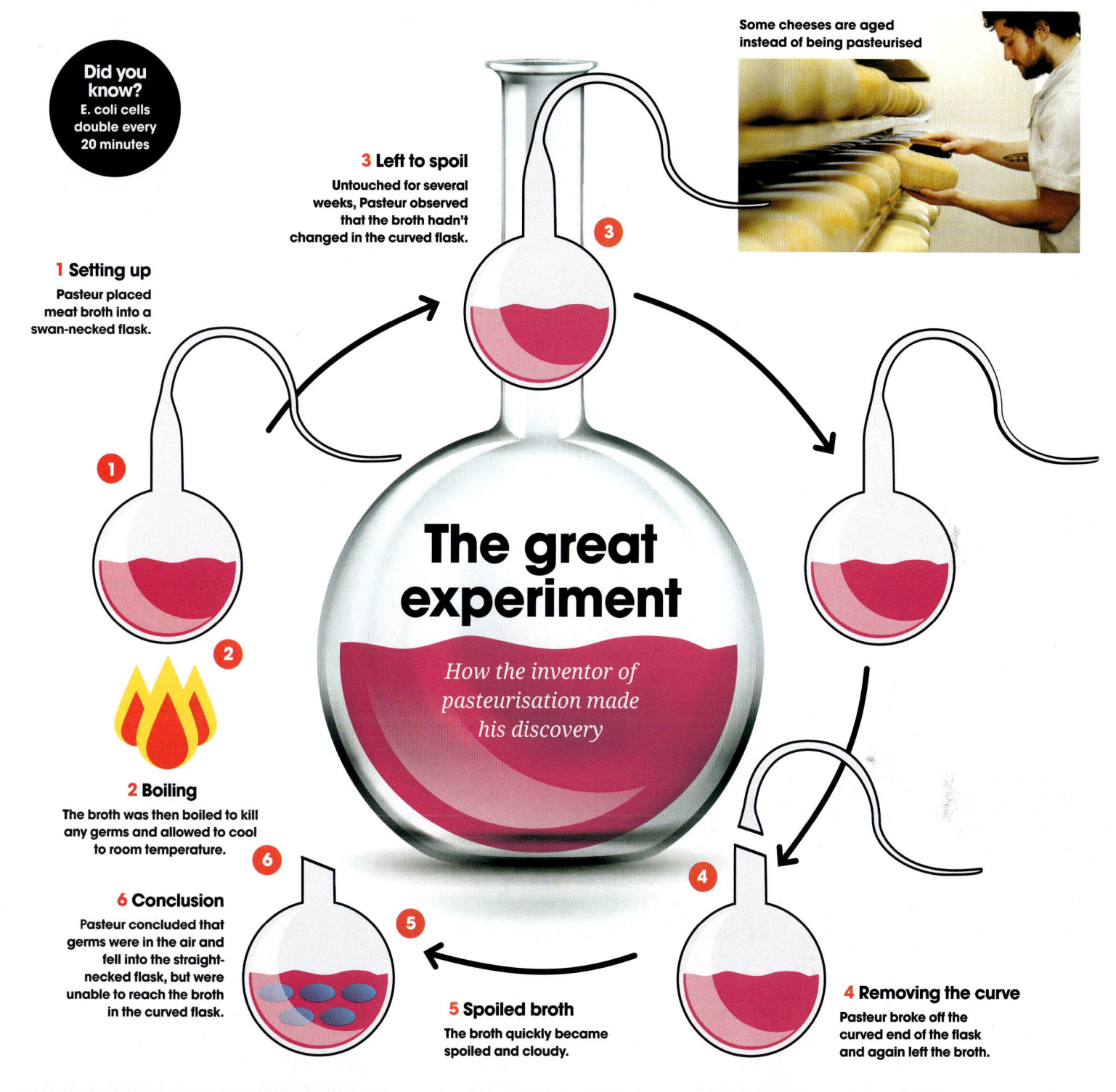

THE POWER OF PASTEURISATION

The great experiment

How the inventor of pasteurisation made his discovery

Did you know?
E. coli cells double every 20 minutes

Some cheeses are aged instead of being pasteurised

1 Setting up
Pasteur placed meat broth into a swan-necked flask.

2 Boiling
The broth was then boiled to kill any germs and allowed to cool to room temperature.

3 Left to spoil
Untouched for several weeks, Pasteur observed that the broth hadn't changed in the curved flask.

4 Removing the curve
Pasteur broke off the curved end of the flask and again left the broth.

5 Spoiled broth
The broth quickly became spoiled and cloudy.

6 Conclusion
Pasteur concluded that germs were in the air and fell into the straight-necked flask, but were unable to reach the broth in the curved flask.

Pasteur the pioneer

The word 'pasteurisation' comes from the man who discovered it in 1865. Louis Pasteur, a French chemist and microbiologist, invented and patented the process after studying the science behind fermentation and attempting to tackle the 'disease' of wine, which was destroying vineyards in France. Having discovered a microbe that converts alcohol into vinegar, called Myoderma aceti, and publishing his 'germ theory' in 1861, Pasteur was tasked by Napoleon III to understand the disease that was turning wine into vinegar. By experimenting with foods that had been exposed to the air following a period of heating and those that had not, Pasteur theorised that germs in the air were responsible for the spoilage of certain foods and wine. He discovered that heating wine to a temperature above 55 degrees Celsius killed microorganisms and extended the wine's life before it spoiled.

A painting of Louis Pasteur experimenting in his laboratory

UNDERSTANDING BIOLOGY

This process doesn't completely sterilise the food product, but does remove enough of the spoilage microorganisms that it's safe to eat and extends its life span. Other than killing harmful pathogens, studies have found that pasteurisation can slightly change the vitamin concentration in some products. In milk the process increases the concentration of vitamin A but decreases the levels of vitamin B12 and E. Some forms of pasteurisation occur long before the food has even been made, such as in eggs. In 2017, the Food Standards Agency updated its advice around eggs, allowing the consumption of raw eggs or slightly cooked eggs produced under British Lion Code. Eggs produced under this code come from chickens that have been vaccinated against Salmonella, therefore reducing the risk of transferring the bacteria. Eggs produced outside of the Lion Code are not suitable for raw eating and should be completely cooked before consuming.

While a majority of foods, such as milk, cheese, nuts and wine, are pasteurised and safe to consume, there are alternatives made from unpasteurised products that are safe to eat. Many soft cheeses are made with unpasteurised or raw milk, but instead of undergoing pasteurisation, this type of cheese is required to be aged at a constant temperature to stunt the growth of harmful bacteria. In Canada the law requires cheese to be aged for at least 60 days at two degrees Celsius before it can be sold.

Typically, the process of pasteurisation requires a constant refrigeration temperature to prevent the growth of spoilage microorganisms, but that's not always the case. Milk that sits in cartons unrefrigerated is known as ultra-high temperature (UHT) milk. This extreme version of pasteurisation involves rapidly heating the milk to temperatures of at least 135 degrees Celsius and then rapidly cooling it to room temperature. This process kills all of the pathogens in the milk, essentially sterilising it, and is packaged in an aseptic container to prevent the growth of any new bacteria. The treated milk can then be stored without the need for refrigeration and can last for months. Although traditional methods of pasteurisation remain the predominant way that dairy products and other perishables are treated, there are emerging non-heat processes, such as High Pressure Processing (HPP) and Pulsed Electric Fields (PEF), that are being explored as an alternative option. PEF foods at ambient temperatures are exposed to pulsing electric fields that neutralise spoilage microorganisms, whereas HPP uses extreme pressure to make spoilage microorganisms inactive and extend the product's shelf life.

Inside a milk pasteurisation plant

Eggs in the United States being pasteurised to remove potential Salmonella

Did you know? 420,000 people die a year from eating contaminated food

Pasteurised products
Reducing our danger from high-risk bacteria

- Fruit juices
- Beer and wine
- Imitation meats
- Nuts
- Flour
- Eggs
- Dairy products
- Honey
- Vinegar
- Crab meat

MICROBIOLOGY

THE POWER OF PASTEURISATION

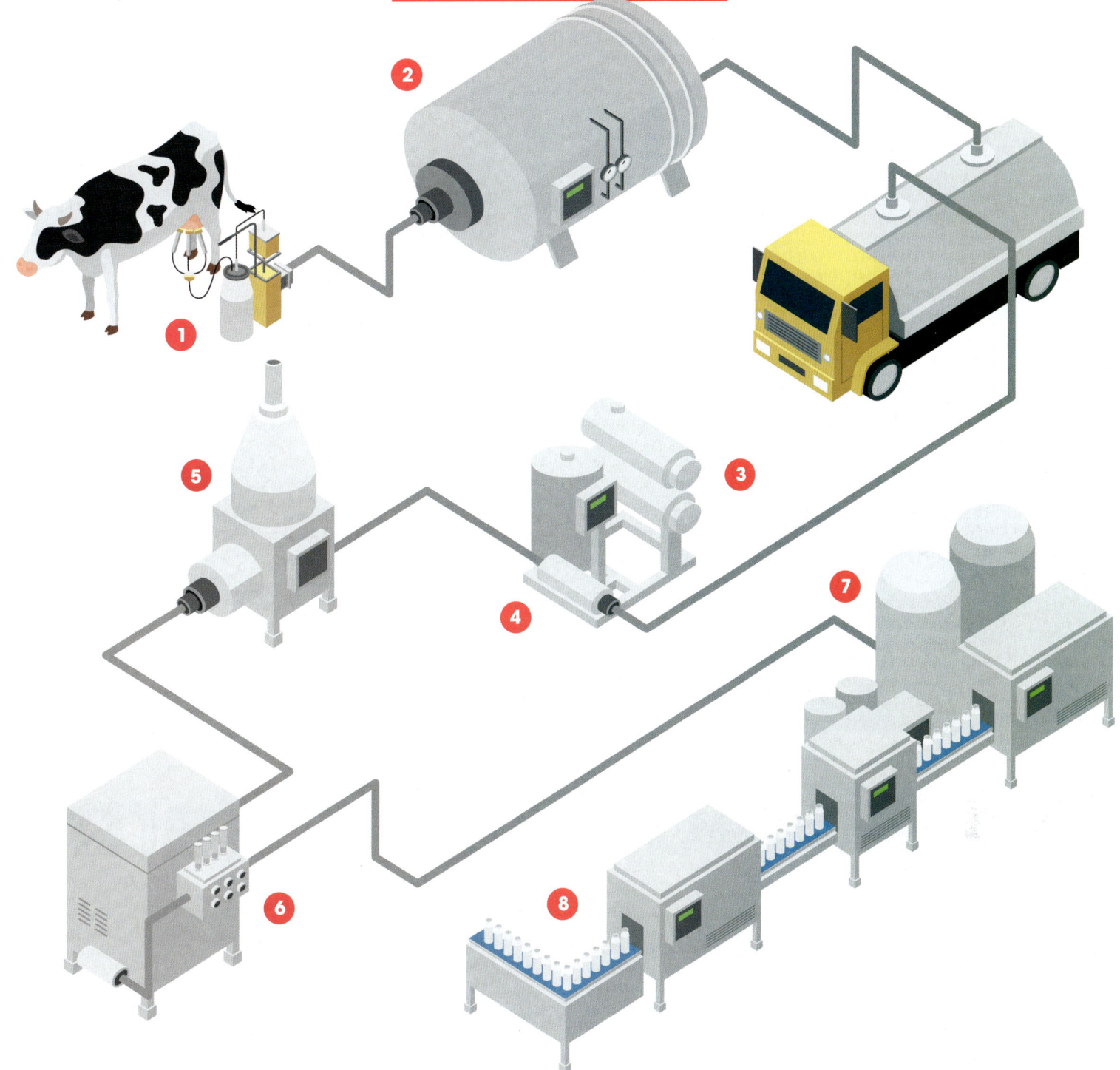

Making milk
The stages that make milk safe to drink

1 Milking
A single dairy cow can produce around 28 litres of milk per day, which is extracted via vacuum cups attached to the udders.

2 Temperature
Milk is typically heated to around 74 degrees Celsius for 20 seconds.

3 Pasteurisation
Milk-filled pipes are surrounded by water at two different temperatures – one to heat the milk and the other to cool it back down to four degrees Celsius.

4 Destruction
The proteins and enzymes in spoilage icroorganisms are denatured and broken down, ultimately killing them.

5 Separation
Milk is spun to separate and remove fat to make the variety of milks on store shelves, such as semi-skimmed.

6 Homogenisation
Pressure is applied to the milk to prevent cream separation and to blend the mixture together.

7 Storage
Pasteurised milk is transported and stored at five degrees Celsius or less.

8 Bottling
Pasteurised milk is packaged into sterilised bottles or cartons and stamped with an expiry date of up to two weeks.

75

UNDERSTANDING BIOLOGY

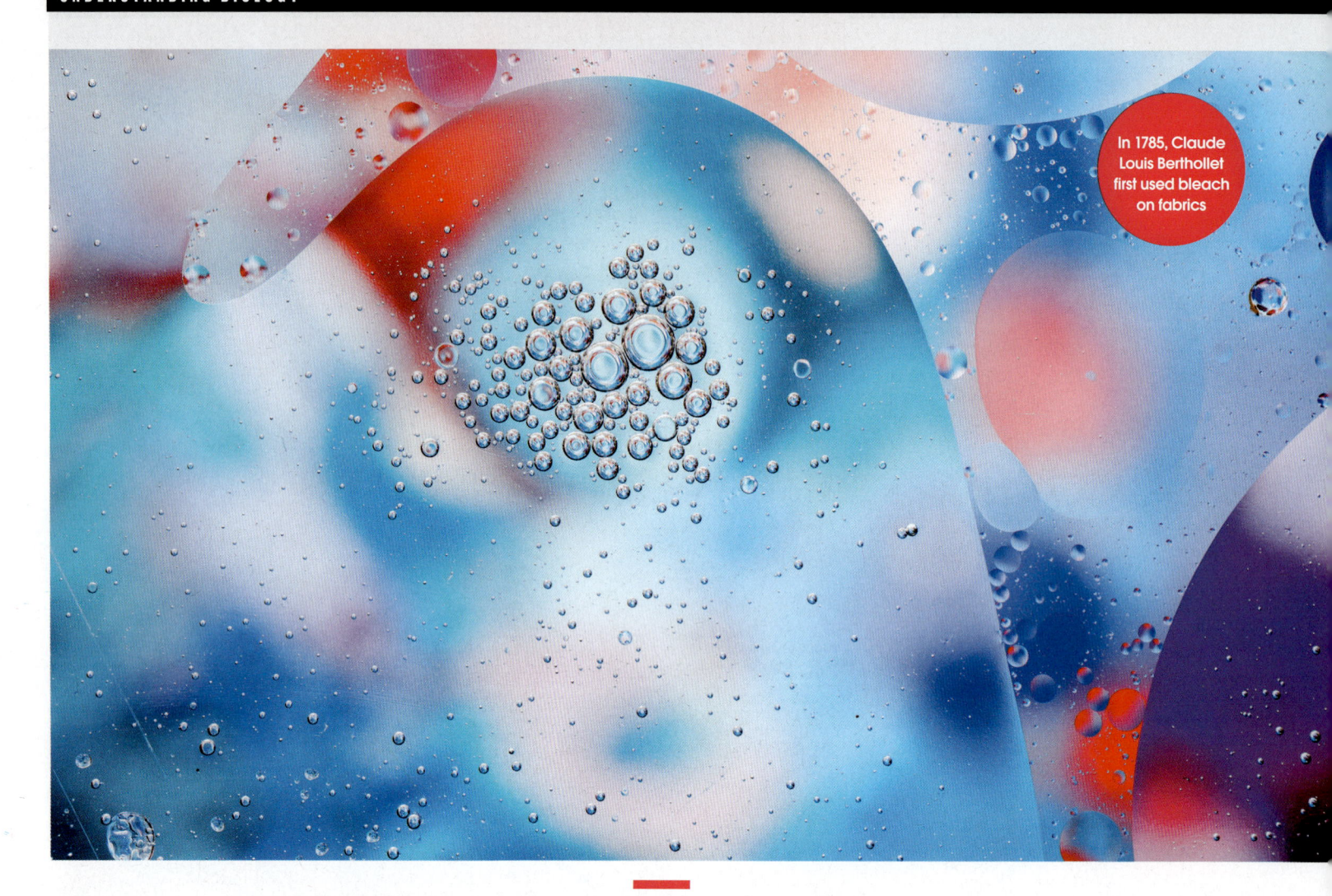

In 1785, Claude Louis Berthollet first used bleach on fabrics

How bleach kills germs

This cleaning cupboard staple tears through proteins, destroying bacteria and viruses in minutes

Household bleach is a bactericide; it kills bacteria on contact. It's one of the cheapest and most powerful microbe busters on the planet. But how does it work? The bleach under your sink is a solution of sodium hypochlorite, which has the chemical formula NaClO. That means that each molecule contains one sodium atom, one chlorine atom and one oxygen atom. In water it produces hypochlorous acid (HOCl), which can steal electrons from other molecules. It uses this ability to attack bacteria and viruses, breaking their proteins apart and splitting them open.

Hypochlorous acid also damages pigment molecules, turning them white, which is where bleach gets its name from. In its pure form, bleach is extremely corrosive. It has a pH of 13, making it just about as alkaline as you can get. Luckily, the bottle under your sink also contains water, lowering the pH to a slightly safer 11.

It doesn't take much bleach to completely sterilise a surface. Your bottle at home is only five per cent bleach, and you need to dilute it even further before you can use it safely. You only need 15 millilitres in 3.8 litres of water to sterilise a table that has been in contact with dirty water. To kill mould, you need a bit more – 240 millilitres in 3.8 litres. Bleach is so reactive that you have to use it within 24 hours of mixing it up or it won't be nearly as effective. It breaks down rapidly into salt and water, especially if it's warm, making it a much less powerful microbe killer. To keep it working at its best, store it in a cool, dark cupboard and only dilute just as much as you need to get the job done.

It takes between 10 and 60 minutes for bleach to kill germs

MICROBIOLOGY

HOW BLEACH KILLS GERMS

How is bleach made?

Bleach is an industrial chemical, and it's produced on an industrial scale. It involves two methods: the chlor-alkali process and the Hooker process. The chlor-alkali process produces the raw materials for bleach production: sodium hydroxide, also known as caustic soda or lye, and chlorine gas. It begins with salty water, also known as brine. The chlor-alkali process uses a technique called electrolysis to break the chemicals in brine apart and remake them into something new. When a current passes through the liquid, it turns water (H2O) and salt (NaCl) into sodium hydroxide (NaOH), chlorine (Cl2) and hydrogen (H2). Two of these three ingredients go on to the next stage, the Hooker process. In the Hooker process, the sodium hydroxide is cooled and the chlorine gas is bubbled through. This produces NaClO, also known as sodium hypochlorite, the active ingredient in household bleach.

Bleach is made from chlorine gas and sodium hydroxide

How bleach kills
Using bleach is a great way to destroy bacteria

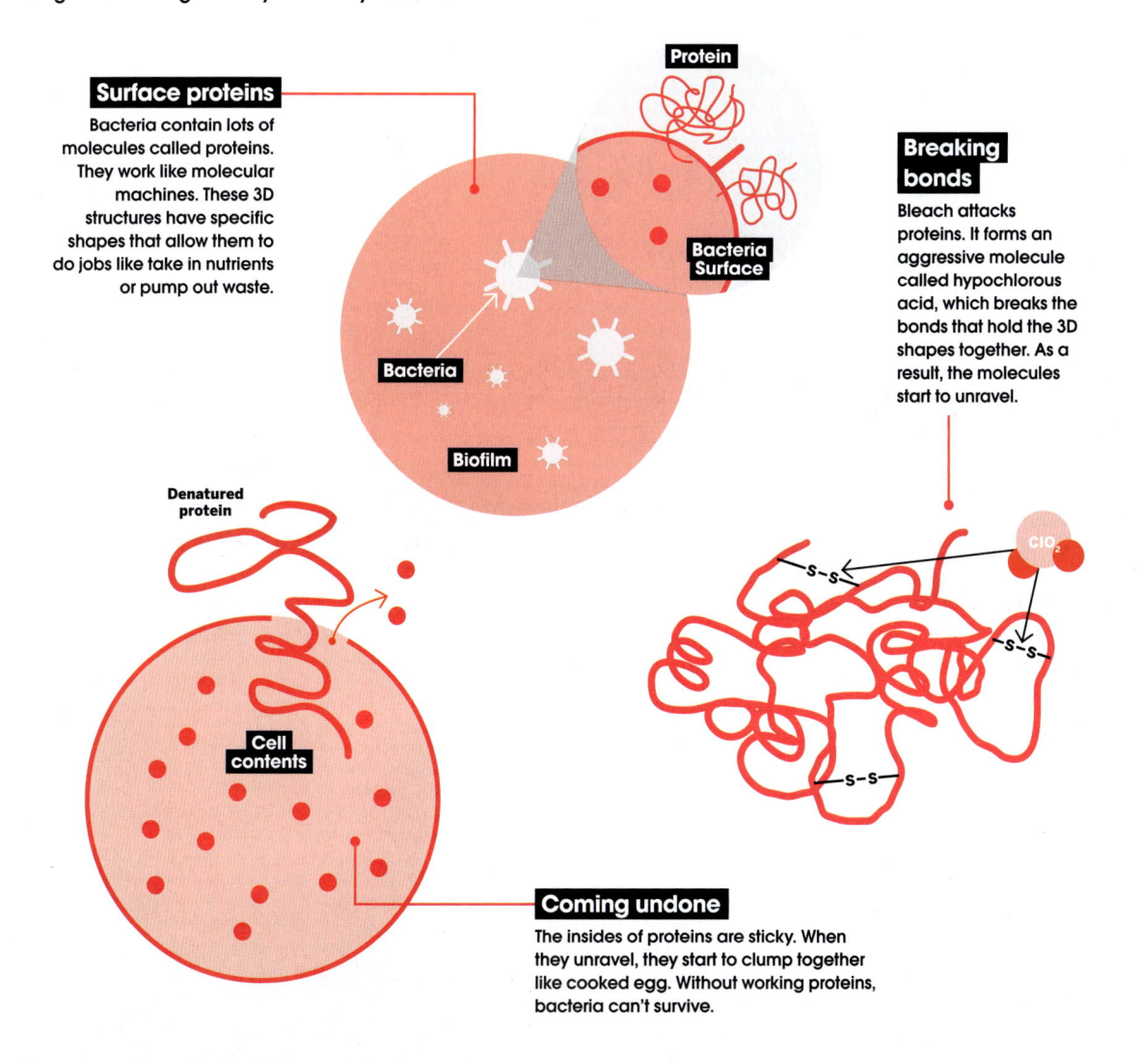

Surface proteins
Bacteria contain lots of molecules called proteins. They work like molecular machines. These 3D structures have specific shapes that allow them to do jobs like take in nutrients or pump out waste.

Protein

Bacteria Surface

Bacteria

Biofilm

Breaking bonds
Bleach attacks proteins. It forms an aggressive molecule called hypochlorous acid, which breaks the bonds that hold the 3D shapes together. As a result, the molecules start to unravel.

Denatured protein

Cell contents

Coming undone
The insides of proteins are sticky. When they unravel, they start to clump together like cooked egg. Without working proteins, bacteria can't survive.

"It doesn't take much bleach to completely sterilise a surface"

5 myths about bleach busted

1 More is better
Bleach is a powerful and dangerous chemical, so more is definitely not better. It only takes one tablespoonful diluted in a gallon of water to kill germs.

2 It causes cancer
While bleach can damage cells and can harm the lungs after severe or long-term exposure, there's no evidence that it can cause cancer in humans.

3 It contains chlorine gas
Bleach contains the chemical sodium hypochlorite, which contains the element chlorine. But the bottle doesn't contain any free chlorine gas.

4 It creates dioxins
Dioxins are toxic ring-shaped chemicals that cause health problems in humans. They're made during the industrial bleaching of paper, but they're not made by household bleach.

5 Drinking it kills coronavirus
Bleach can kill the SARS-CoV-2 coronavirus on surfaces, but it doesn't work on or inside the body. It causes serious harm to human cells.

UNDERSTANDING BIOLOGY

Lab-grown meat explained

Would you eat chicken or beef that's been grown in a laboratory?

Could lab-grown meat be the compromise that settles the conflict between meat-eaters and vegetarians? For more than 2.5 million years humans have eaten the flesh of other animals. During this time, the world's agriculture industry has boomed, with 26 per cent of Earth's ice-free land dedicated to livestock, which comes at a cost to the environment. On average, the livestock industry is responsible for around 18 per cent of global greenhouse gas emissions.

To help minimise environmental impact and reduce the space needed to house livestock, scientists have started growing food far from the fields, inside their laboratories. Scientists care for a collection of tiny cells that have been extracted from the genuine article without the need to slaughter a living animal.

A small sample of stem cells, typically taken from animal muscle via a small biopsy under local anaesthetic, is collected and placed in a culture to grow. Stem cells act like the understudies of the body's other cells: when stimulated, stem cells can develop into specialised cell types, such as muscle and fat cells. To grow meat, which is essentially a collection of fat and muscle cells, scientists expose the extracted stem cells to amino acids and carbohydrates that trigger their ability to change. The changing stem cells are left to grow and multiply until a heap of muscle fibres has developed and the new meat is plentiful enough to use in meat products, such as burgers.

So far, only a handful of animal species have had their cells turned into this novel form of meat, including cows, chickens and pigs. However, some companies plan to expand into more exotic meats, such as kangaroo, alligator and ostrich. Currently, lab-grown meat isn't widely available. The first country to allow the sale of cultured meat was Singapore in 2020. In 2023, American companies such as the San Francisco start-up Upside Foods were given the green light to sell their cultured meat products in the US.

1.5 billion cows are eaten globally each year

Mammoth meatballs

So far, lab-grown meat has been limited to cows, chickens and pigs. But what if you could feast on the meat of animals long extinct, like the mighty mammoth? An Australian company called Vow has done just that. Combining the DNA of the extinct woolly mammoth (Mammuthus primigenius) and its closest living relative, the African elephant (Loxodonta), Vow served up the first mammoth meatball in March 2023. Although the meatball looked tasty enough to eat, it's yet to pass anyone's lips. Fears over the safety of the meat and the possible negative immune response of the human body mean that the taste of mammoth meat remains a mystery.

The mammoth is back, but in meatball form

MICROBIOLOGY

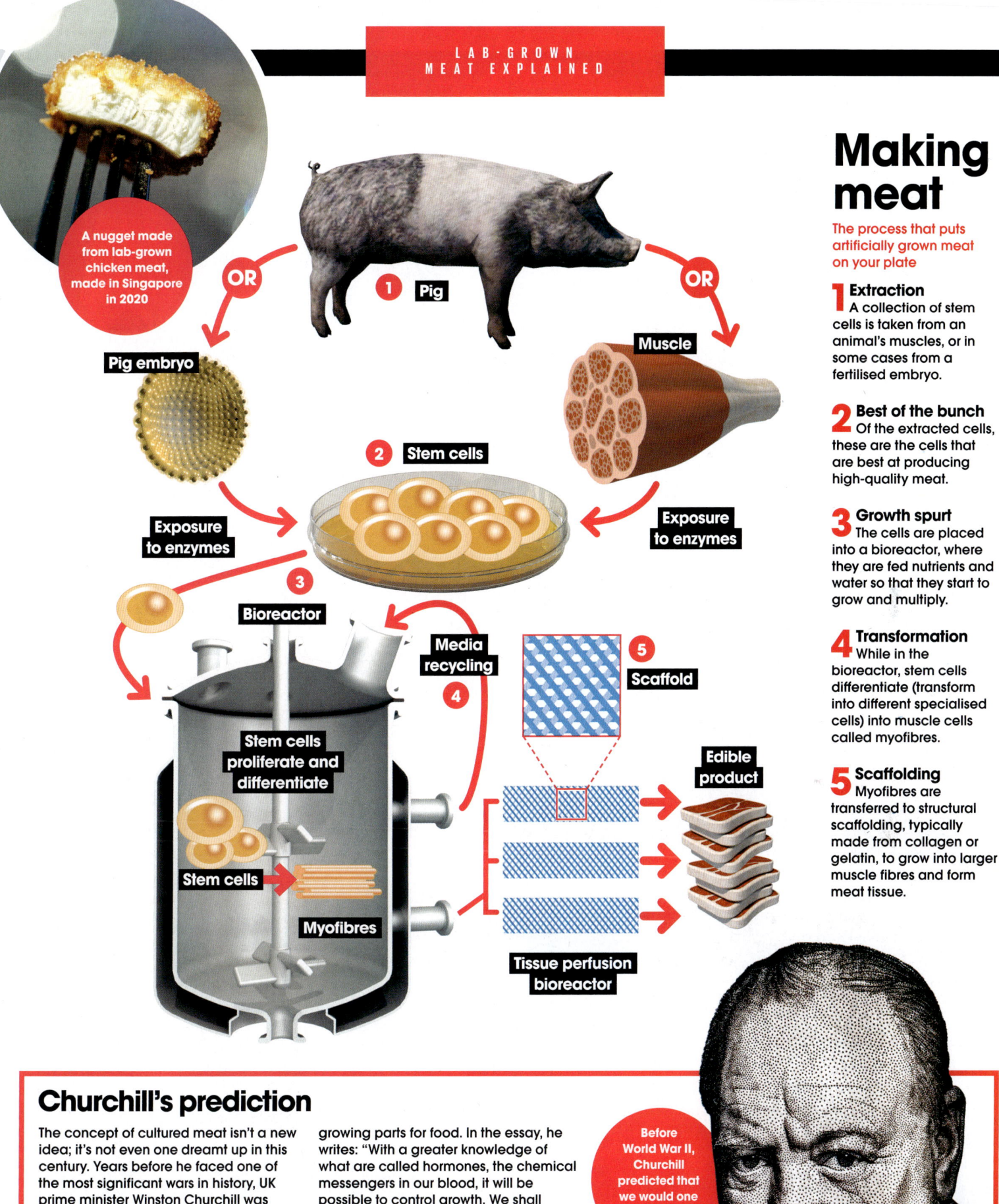

LAB-GROWN MEAT EXPLAINED

Making meat

The process that puts artificially grown meat on your plate

1 Extraction
A collection of stem cells is taken from an animal's muscles, or in some cases from a fertilised embryo.

2 Best of the bunch
Of the extracted cells, these are the cells that are best at producing high-quality meat.

3 Growth spurt
The cells are placed into a bioreactor, where they are fed nutrients and water so that they start to grow and multiply.

4 Transformation
While in the bioreactor, stem cells differentiate (transform into different specialised cells) into muscle cells called myofibres.

5 Scaffolding
Myofibres are transferred to structural scaffolding, typically made from collagen or gelatin, to grow into larger muscle fibres and form meat tissue.

A nugget made from lab-grown chicken meat, made in Singapore in 2020

1 Pig — OR — Pig embryo / Muscle
2 Stem cells (Exposure to enzymes)
3 Bioreactor — Stem cells proliferate and differentiate — Myofibres
4 Media recycling
5 Scaffold → Edible product
Tissue perfusion bioreactor

Churchill's prediction

The concept of cultured meat isn't a new idea; it's not even one dreamt up in this century. Years before he faced one of the most significant wars in history, UK prime minister Winston Churchill was known for writing predictive essays on the state of humankind. In one such essay, entitled *Fifty Years Hence*, which was published in *The Strand Magazine* in 1931, Churchill predicted that as scientific understanding grew, we may no longer need the whole animal, only growing parts for food. In the essay, he writes: "With a greater knowledge of what are called hormones, the chemical messengers in our blood, it will be possible to control growth. We shall escape the absurdity of growing a whole chicken to eat the breast or wing by growing these parts separately under a suitable medium."

It took 82 years, but in 2013 the first lab-grown meat product, a burger, was created by scientists in the Netherlands.

Before World War II, Churchill predicted that we would one day culture our own meat

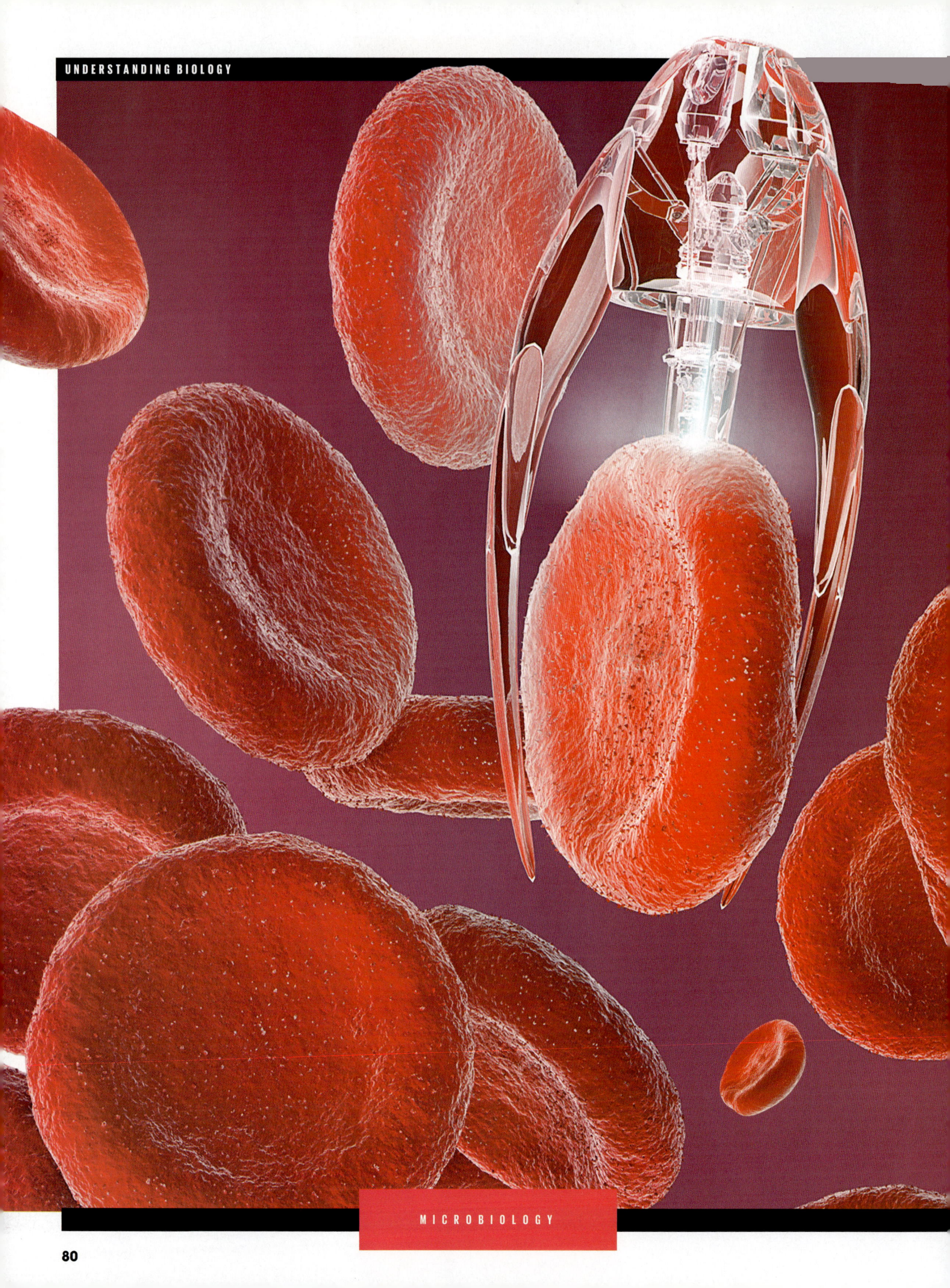

UNDERSTANDING BIOLOGY

MICROBIOLOGY

WHAT IS NANOTECH?

What is Nanotech?

Enter the invisible world of tiny machines and materials, where everything measures 0.0001 millimetres or less

The smallest object visible to the naked eye is a human egg cell. It measures just 0.1 millimetres in length. Beyond that limit, the world is completely invisible. Nanotechnology is the field of science and engineering at a scale of one to a hundred nanometres. That's a thousand to a hundred thousand times smaller than the smallest thing we can see.

Particles at this tiny scale behave completely differently to the full-size structures we are used to. Gold changes colour from yellow to purple and becomes liquid at room temperature. Carbon transforms into an extraordinary electrical conductor. And copper gains the ability to kill bacteria. Discovering this invisible world and harnessing its power is the domain of nanotechnology.

UNDERSTANDING BIOLOGY

The first person to measure and name a nanoparticle was Richard Zsigmondy. Awarded the Nobel Prize in Chemistry in 1925 for his efforts, he was fascinated by ceramics and glass. Their stunning colours were the result of nanoscale particles known as colloidal gold, which reflect light in unusual ways. But it wasn't until the 1950s that people started to experiment with using nanoparticles like these in different ways. During the Cold War, another Nobel Prize winner, Richard Feynman, was investigating the possibility of science on an atomic scale. He published a paper entitled There's Plenty of Room at the Bottom, inviting scientists to enter a new field of physics. Feynman wanted to miniaturise computers and create machines that could assemble molecules atom by atom. And he dreamed of a day when you could 'swallow your surgeon', delivering a life-saving robot into your body.

His ideas might sound like science fiction, but nature had already proven they were possible. Inside every single cell there are millions of molecules, each measuring just fractions of a millimetre across. These particles can perform both mechanical and chemical work; they self-assemble and self-heal, and they also store and exchange information. They are nature's nanomachines – all we have to do is recreate them.

In the 1980s, two powerful new microscopes revealed the nanoworld in all its minute detail. The scanning tunnelling microscope used a very fine wire and an electric current to map even the thinnest of materials. At nanoscale, electrons behave like waves, allowing them to pass through solid objects using a technique called tunnelling.

The atomic force microscope worked in a similar way, but used a silicone tip and a laser to trace the surface of a sample. It could tell researchers about the magnetic, electrical, chemical and physical properties of materials they could never hope to see with their own eyes.

By the 1990s the field had exploded. Researchers observed carbon nanotubes for the first time – just a single atom thick, they were stronger than steel. In the years that followed, scientists developed the ability to fold DNA like paper, and they discovered that it was possible to guide nanoparticles inside the body like homing missiles. Now, with science advancing at lightning speed, Feynman's dreams are closer than ever.

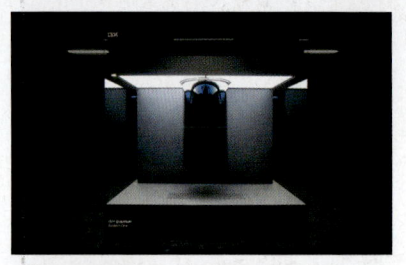

Did you know? Nanofibres can filter pollutants from indoor air

Factories make graphene by forcing ions between layers of graphite to peel them apart

IBM invented the world's first circuit-based commercial quantum computer

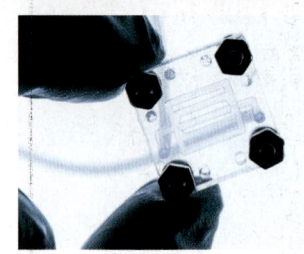

Organs-on-chips use nanotechnology to simulate parts of the human body

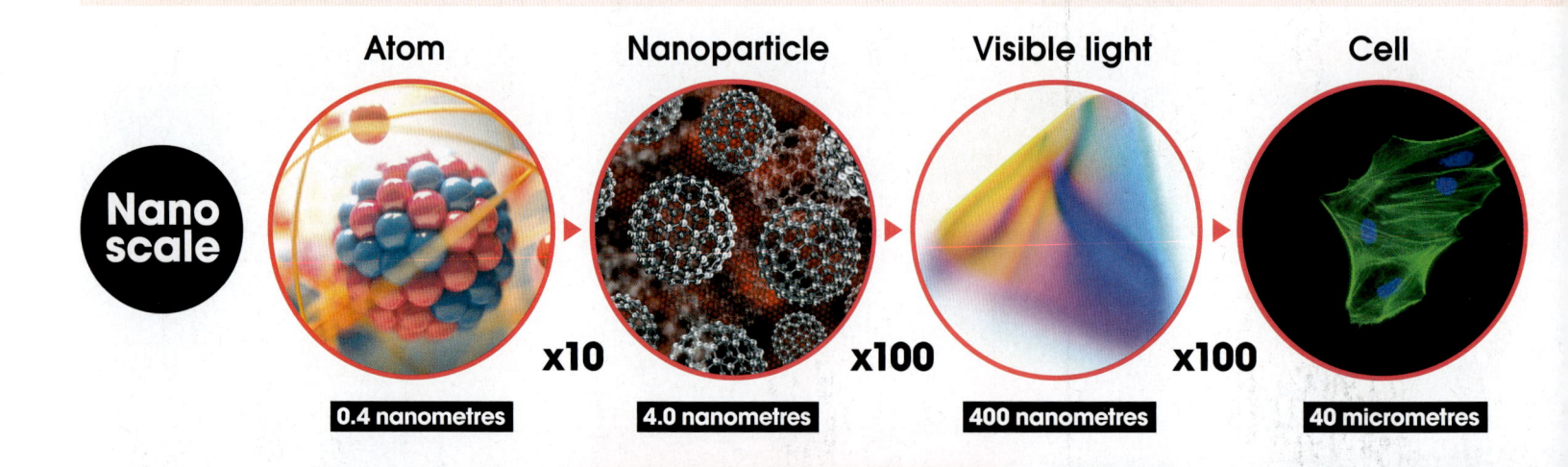

> "It wasn't until the 1950s that people started to experiment with using nanoparticles in different ways"

Nano scale

Atom	Nanoparticle	Visible light	Cell
0.4 nanometres	4.0 nanometres	400 nanometres	40 micrometres
	×10	×100	×100

MICROBIOLOGY

WHAT IS NANOTECH?

Types of nanomaterials
Each of these small-scale particles has the potential to change the world

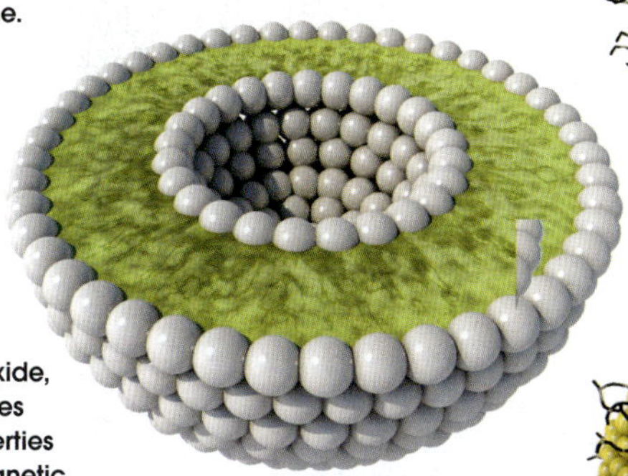

7 Liposome
These little spheres can carry cargo into living cells. Made from cholesterol and fats called phospholipids, they could help deliver drugs that wouldn't normally be able to cross the cell membrane.

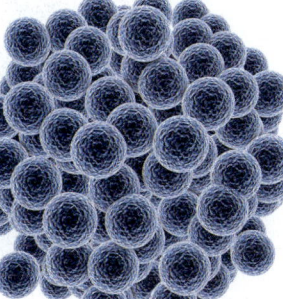

6 Magnetic
Made from iron oxide, these nanoparticles have all the properties of a full-sized magnetic material. They show promise in medical diagnosis and treatment because magnets can guide them into position.

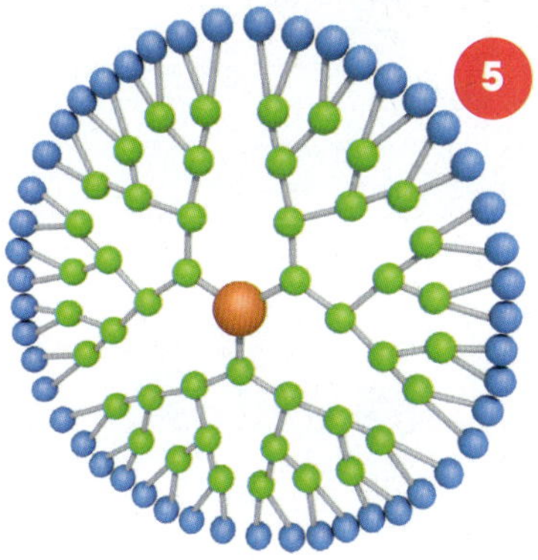

5 Dendrimer
These symmetrical nanoparticles have arms that look a bit like the branches of trees. They can carry molecules inside their cores, making them a possible delivery vehicle for cancer treatments.

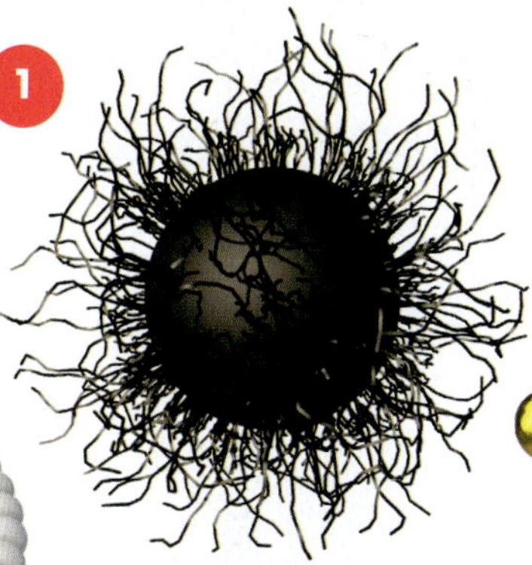

1 Micelle
These self-assembling structures have an outer water-loving shell and an inner water-hating core. They can help stubborn chemicals dissolve by shielding them from water, potentially aiding drug delivery.

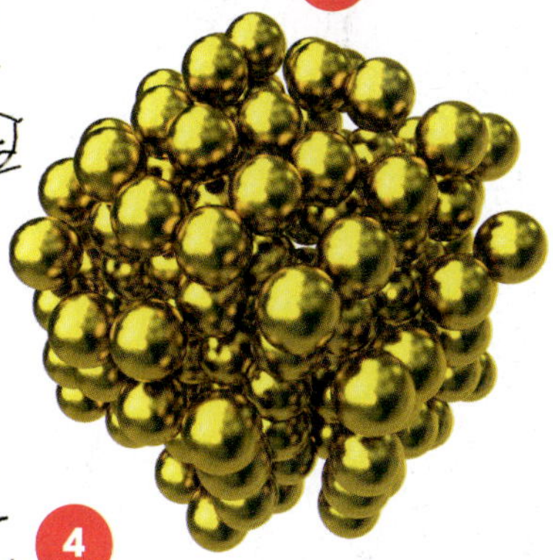

2 Gold
Made from an inert precious metal, these nanoparticles show promise in medical treatment. They can carry chemicals and genetic material to precise locations inside the body without causing any harm.

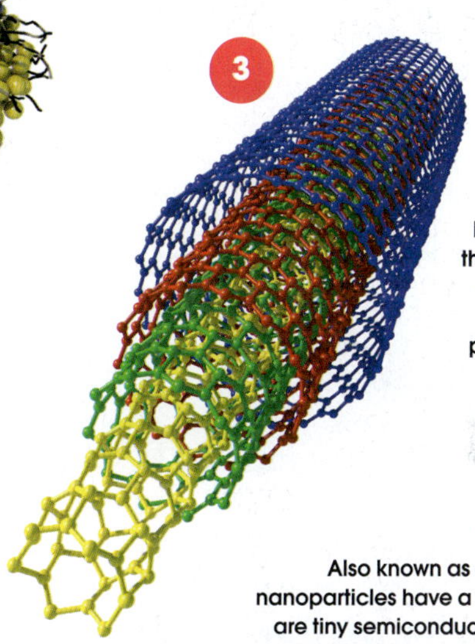

3 Carbon nanotube
These large-scale nanoparticles form hollow rods one atom thick and with a tensile strength greater than steel. They have potential applications in everything from display screens to artificial limbs.

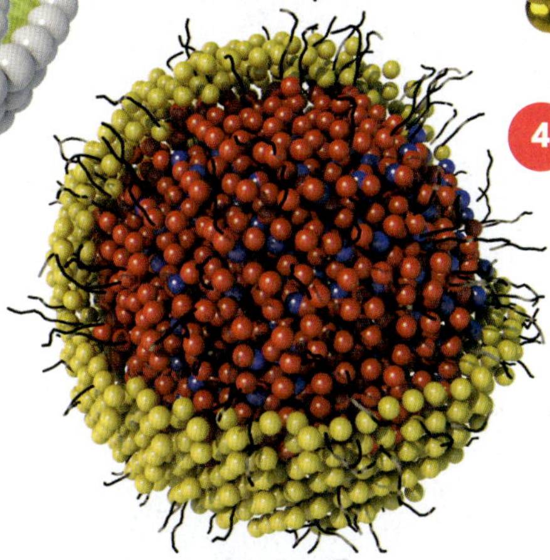

4 Quantum dot
Also known as artificial atoms, these nanoparticles have a crystal structure. They are tiny semiconductors with the ability to absorb photons and emit light, making them attractive for display technology.

Ant	Rabbit	Aeroplane	Town
x100	x100	x100	x100
4.0 millimetres	40 centimetres	40 metres	4,000 metres

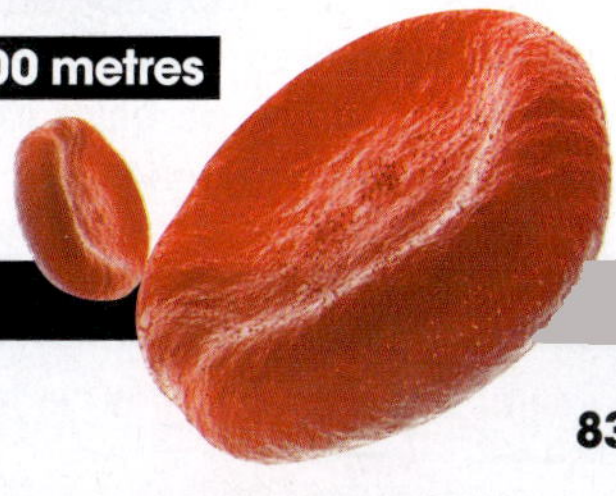

© Getty / Wiki: Jweiss20 / Illustration by Adrian Mann

83

UNDERSTANDING BIOLOGY

Cell-sized robots could revolutionise medical diagnosis and treatment

Introducing nanobots

Nanobots are robots on a molecular scale. Smaller than a grain of sand, these programmable particles take inspiration from biology. Inside every living cell are tiny machines that are surprisingly like the machines we use in our everyday lives. There are pumps, motors, switches and even clocks, all working together to perform complex work in miniature. Researchers have taken inspiration from these machines to create brand-new technology from tiny component parts. Made from proteins, fat and even DNA, nanobots have the potential to perform tasks like delivering drugs, killing bacteria and inactivating toxins. Some of the most promising use a design pattern called DNA origami. Made from repeating units of genetic code, these tiny particles can self-assemble into geometric shapes. They can form hollow cubes, rod-like scaffolding and even gears. In the future, particles like these could recreate mechanical machines on a nanoscale.

Wearable nanopatches

Skin-mounted sensors can detect deadly gases, monitor the body or deliver life-saving drugs

1 Sensor array
Miniature sensors can detect anything from touch and temperature to chemicals in the air.

2 Flexible materials
Nanomaterials mixed with fabrics or polymers create flexible sheets that can move with the body.

3 Self-powered
Nanogenerators harvest electricity from friction, heat, sound, vibration and even wind or blood flow.

4 Drug delivery
Patch-mounted pumps can deliver life-saving treatments, like insulin for diabetes.

MICROBIOLOGY

WHAT IS NANOTECH?

Fighting cancer with nanobots

Miniature robots from Arizona State University destroy tumours by cutting off their blood supply

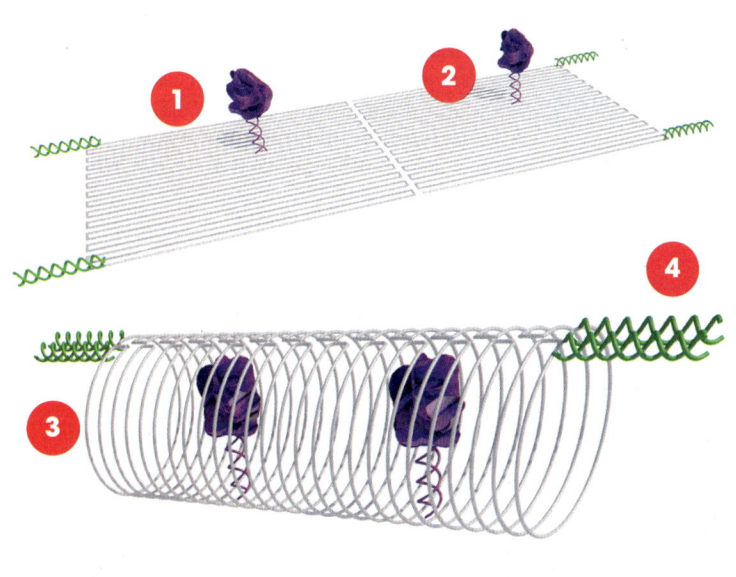

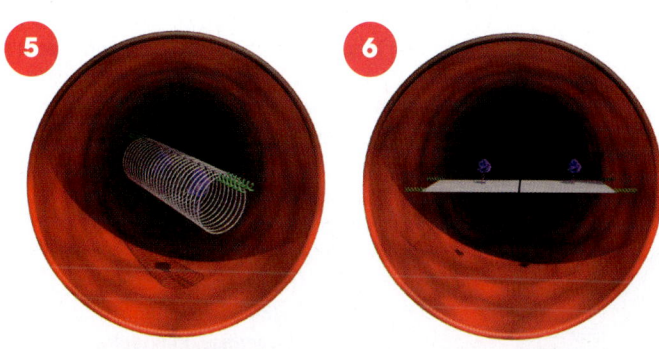

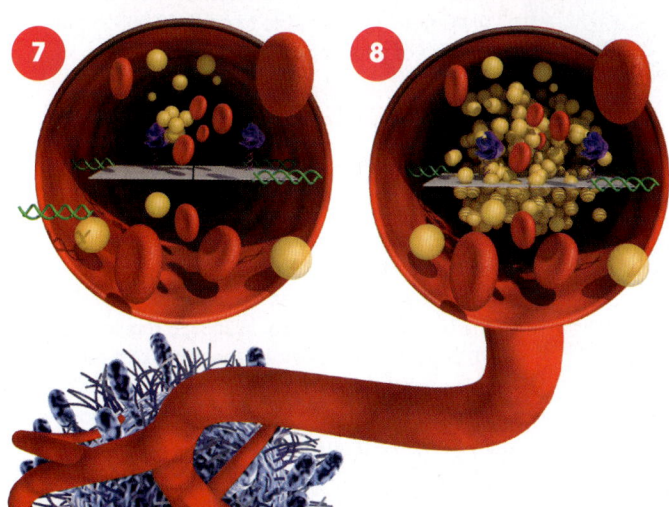

1 DNA sheets
Flat, rectangular sheets of DNA form the basic scaffolding of the nanobots.

2 Thrombin
The sheets carry a protein called thrombin, which causes blood to clot.

3 Origami
The DNA folds over to form a tube around the thrombin.

4 Homing signal
Fragments of DNA at the edges of the sheet guide the nanobots to molecules found on the surface of cancer cells.

5 Inside the body
Each nanobot measures just 60 nanometres in length, allowing it to fit easily into tiny blood vessels.

6 Trojan horse
When the nanobots reach the tumour they uncurl, revealing the thrombin proteins hidden inside.

7 Clotting begins
The blood starts to clot around the nanobots, sticking together to form a solid plug.

8 Tissue damage
Within 24 hours, the clot blocks the blood supply to the tumour, killing the cancer cells.

Nanodentistry

Nanobots have the potential to revolutionise dentistry. Performing surgery in the mouth is difficult. There isn't a lot of room to move around, and there are lots of vital tissues, nerves and blood vessels close together. In the future, nanobots could travel through the gums and deliver anaesthetic directly to an individual tooth, without the need for needles. They could use nanoparticles made from the bone mineral hydroxyapatite to repair cavities and jaw damage. And they could even destroy harmful bacteria. Future nanodentistry could also include high-tech mouthwash or toothpaste to deliver a protective nanocoating to shield teeth from plaque and tartar.

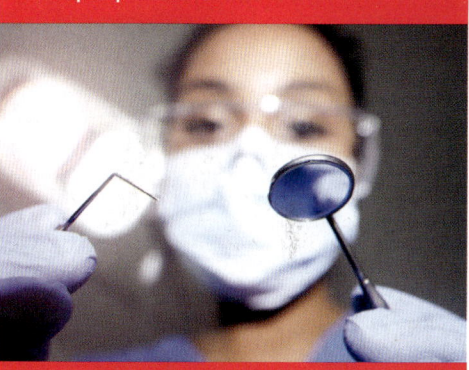

Nanotechnology could make dentistry painless and precise in the future

Electrospun nanofibres could make ultra-fine virus filters for masks

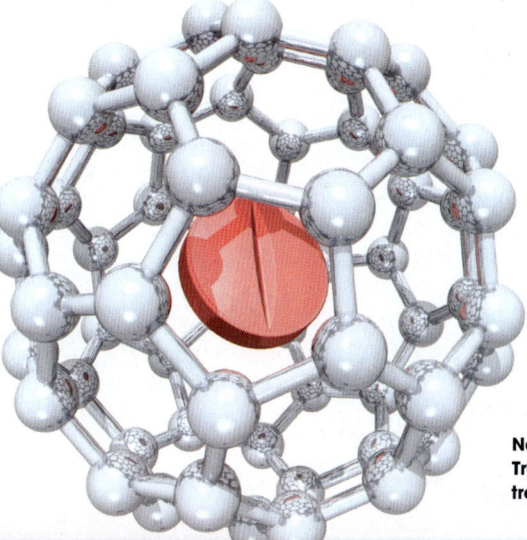

Nanoparticles work like Trojan horses, carrying treatments into cells

UNDERSTANDING BIOLOGY

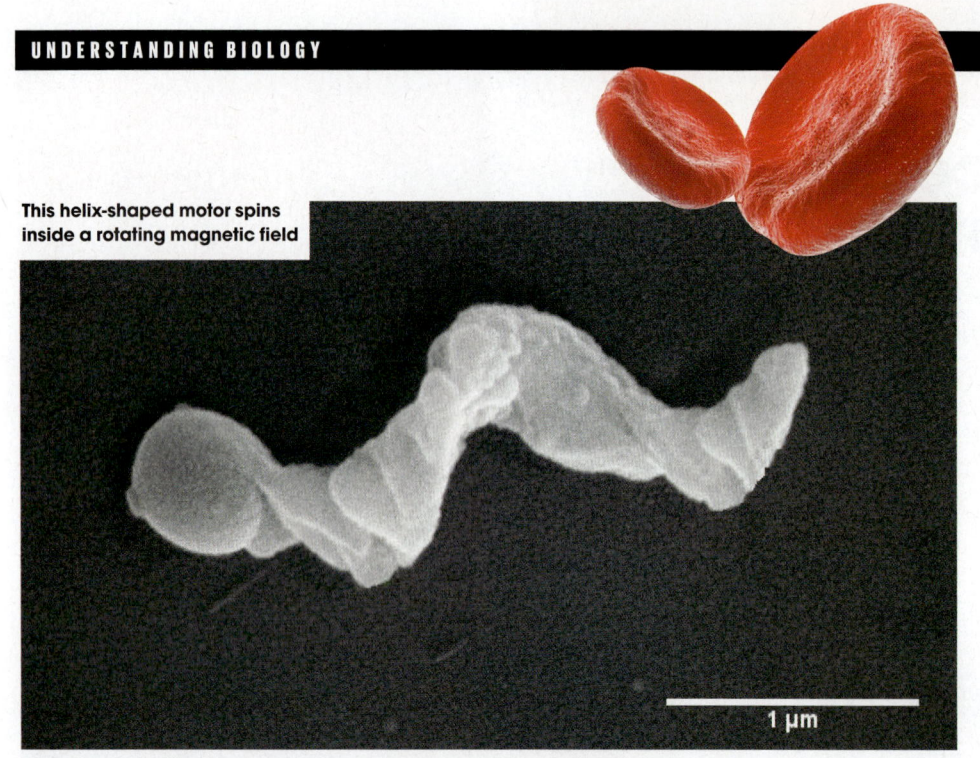

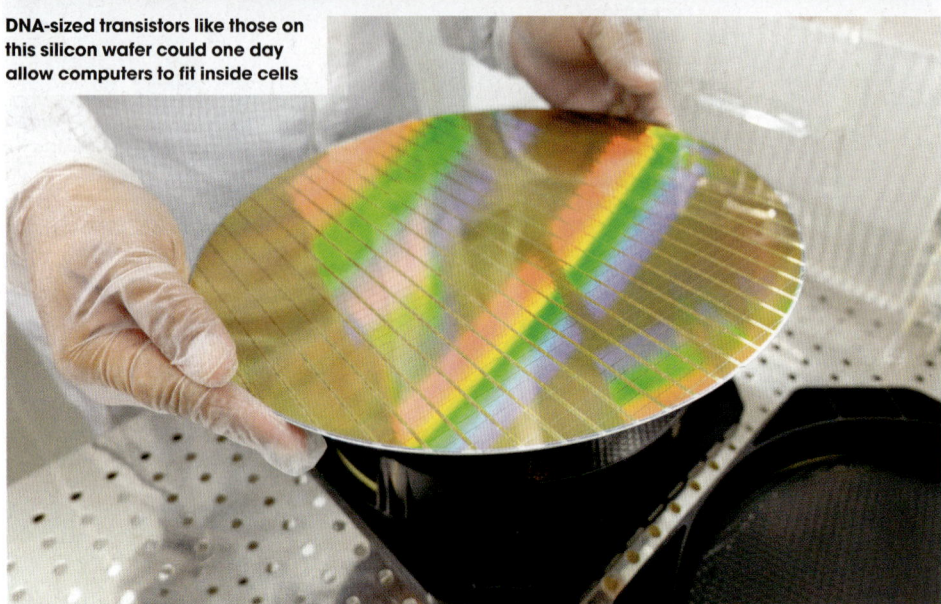

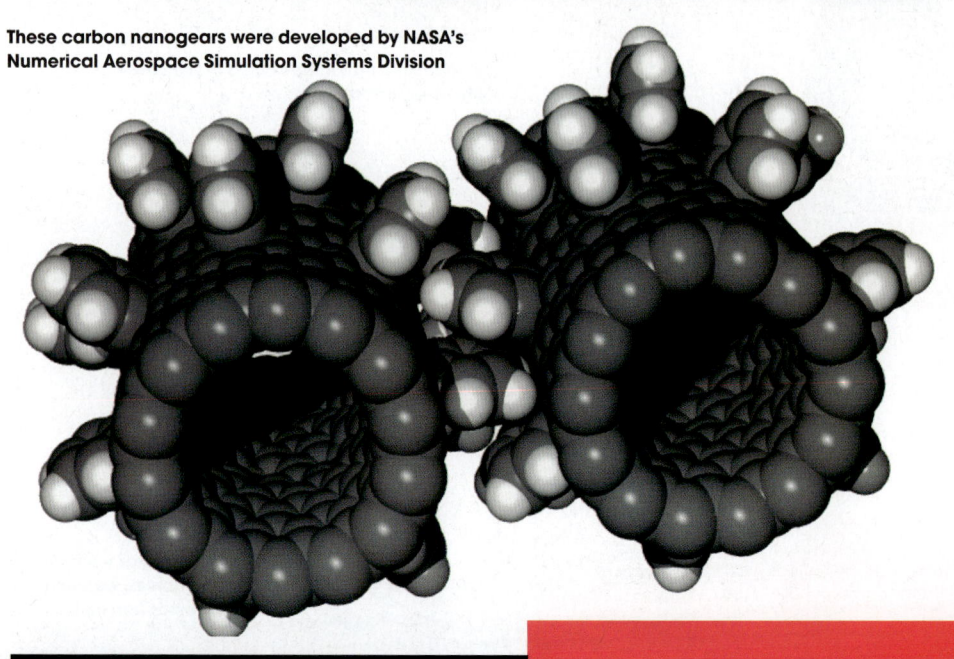

This helix-shaped motor spins inside a rotating magnetic field

DNA-sized transistors like those on this silicon wafer could one day allow computers to fit inside cells

These carbon nanogears were developed by NASA's Numerical Aerospace Simulation Systems Division

Magnetic nanobots

Magnetic nanoparticles respond to external magnetic fields. This makes it possible to guide them through space, opening the way for magnet-driven nanobots. Researchers have found that applying different types of magnetic fields can make the tiny robots move in different ways. A rotating magnetic field can make a helical robot spin, winding forward like spring. An oscillating magnetic field can make a nanobot undulate, curling up and down like a fish. And a stepping magnetic field can make it stop and start like a switch, allowing a nanobot's limbs to perform rowing movements. It's also possible to combine magnetic control with other types of input, like light or ultrasound. This could one day allow much finer control of a nanobot's movement, allowing it to perform highly complex and extremely delicate tasks, like surgery.

Ultra-powerful computers

Computers think in zeros and ones. They contain millions of tiny switches, called transistors, that can be either on (one) or off (zero). The more transistors a computer has, the faster it can think. For the past few decades, engineers have been working hard to make this technology as small as possible. As a result, the number of transistors that can fit on a computer chip has roughly doubled every two years – a phenomenon known as Moore's law. Your average computer now contains transistors no bigger than 14 nanometres across, while the most advanced contain transistors that measure just seven nanometres. In 2022, IBM unveiled transistors that measure just two nanometres, thinner than a strand of DNA, making it possible to pack 50 billion onto a chip the size of your fingernail.

Fullerene machines

In 2016, three chemists received the Nobel Prize for developing the world's smallest machines. They had invented chains, axles and motors on a molecular scale. These advances led to the development of the world's first nanocar in 2005, a moving vehicle 20,000 times smaller than the width of a human hair. The car had wheels made from buckyballs and axles that moved freely, allowing the whole structure to roll. By 2017, researchers were literally racing to develop the next generation of nanoautomobiles. The world's first nanocar race was held in France, using a scanning tunnelling microscope to propel six of the world's smallest cars towards the finish line. These experiments are good for more than just entertainment, however. They could one day lead to programmable machines capable of performing mechanical work on a miniature scale.

MICROBIOLOGY

WHAT IS NANOTECH?

5 Facts
Nanotech that already exists

1 Computer processors
The processor that powers your computer or your phone contains hundreds of tiny switches called transistors. Each one measures less than ten nanometres across.

2 Quantum displays
Ultra-high-definition QLED screens contain tiny nanocrystals called quantum dots – the Q in QLED stands for quantum. They contain nanoscale semiconductors that emit different coloured light depending on their size.

3 Molecular printers
It's now possible to 3D print at nanoscale using an AI-powered scanning tunnelling microscope. The powerful piece of kit can pick up and lay down molecular building blocks.

4 Flexible screens
Advances in nanotechnology have made it possible to bend, roll and even fold electronic screens. They use flexible plastic polymers and ultrafine conductive materials like graphene or silver nanowires.

5 Self-healing plastic
New types of plastic combine flexible paint-like materials with tough nanoscale polymers to allow tears and breaks to self-heal. The process even works under water.

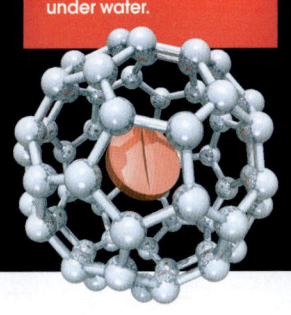

Flexible nanotech

Sheet materials just one atom thick have the potential to become anything from bendy screens to e-tattoos

- Smart glasses
- Wearables
- Smart phones
- E-tattoos
- Smart watches

> "A rotating magnetic field can make a helical robot spin"

Graphene has the potential to make ultrafast electrical circuits

Ferrofluids have nanoscale magnetic particles that form peaks and valleys when exposed to magnetic fields

87

UNDERSTANDING BIOLOGY

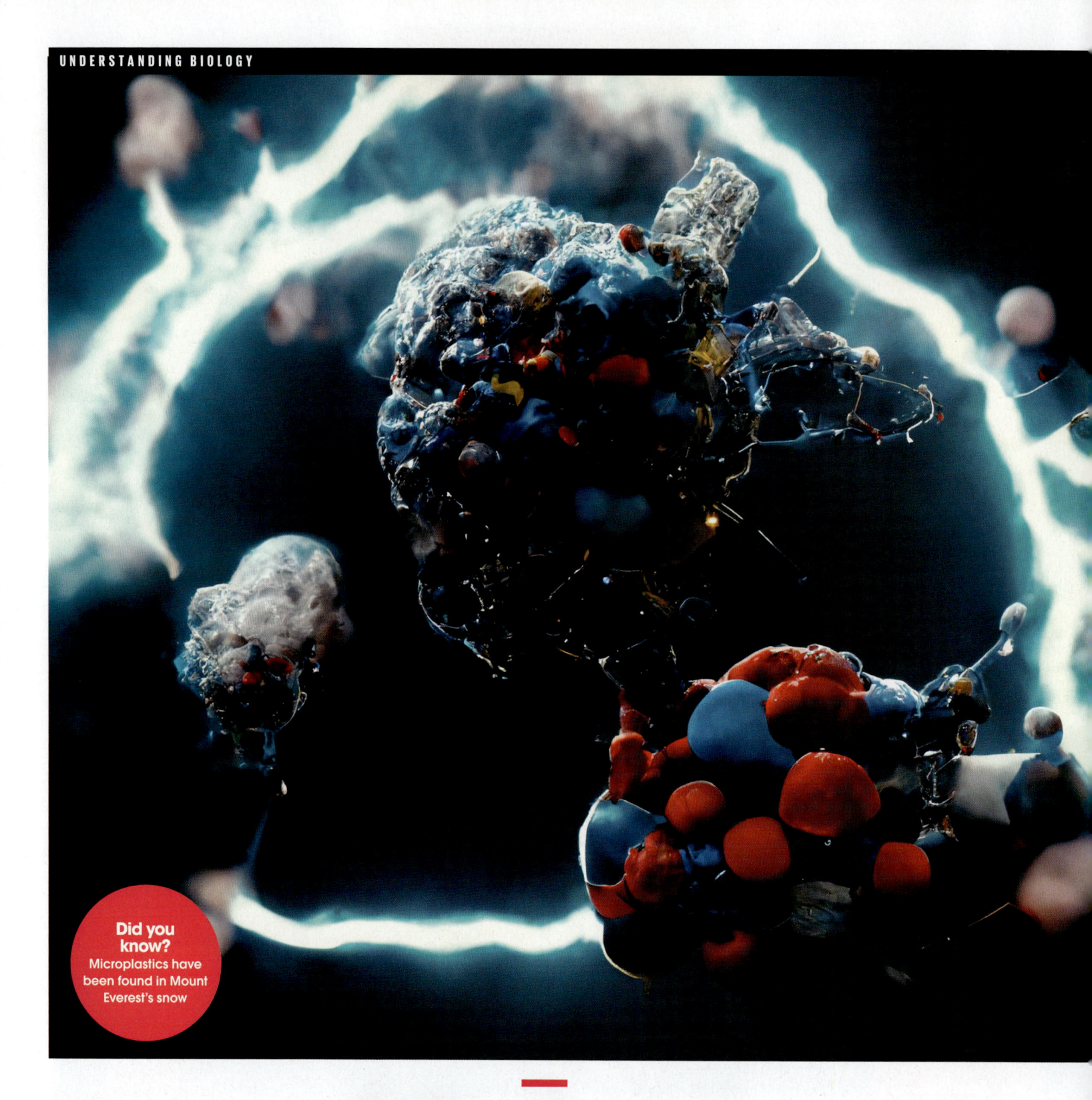

Did you know?
Microplastics have been found in Mount Everest's snow

What are microplastics?

Take a closer look at these miniscule pollutants and how they affect all life

MICROBIOLOGY

WHAT ARE MICROPLASTICS?

Plastics are made from materials such as oil and plant minerals found naturally on Earth, but the end product itself is entirely unnatural. This synthetic product can now be found in every corner of the world, in remote places like the deep ocean and Antarctica, and even inside our bodies. This is because plastic doesn't fully decompose.

Every piece of plastic produced still exists in some form, even if it's out of sight. In most instances, plastic products have been broken down into microplastics, tiny fragments that have been reduced to a length of five millimetres or less, too small to see with the naked eye. The first evidence of microplastic pollution was discovered in the oceans, where scientists have estimated that there are 24.4 trillion microplastic particles in the upper oceans alone.

These tiny pieces of plastic are divided into primary and secondary microplastics. Primary are those that are intentionally produced by manufacturers in this form, while the latter are broken down in the environment from larger objects.

> "Every piece of plastic produced still exists in some form, even if it's out of sight"

5 types of microplastics

1 Fibres
These thin, thread-like plastics come from fleece clothing and cigarette butts. One of the main ways these fibres enter lakes is through washing machines. One fleece jacket can release 2,000 during a single wash.

2 Microbeads
These spheres have a diameter less than one millimetre. They are often used to add exfoliating properties in soaps but are difficult to filter out of waste water due to their small size. Microbeads are banned in many countries.

3 Fragments
When a larger piece of plastic breaks down over time, it becomes smaller microplastic fragments. Usually this is a result of exposure to the Sun's ultraviolet radiation.

4 Nurdles
Small pellets that are manufactured and used to make larger products. The small beads are melted and shaped into moulds of plastic products. Sometimes these escape into lakes and oceans during transportation.

5 Foam
This is the Styrofoam often used as packing protection in delivery boxes. This soft product can be broken down easily into microplastics.

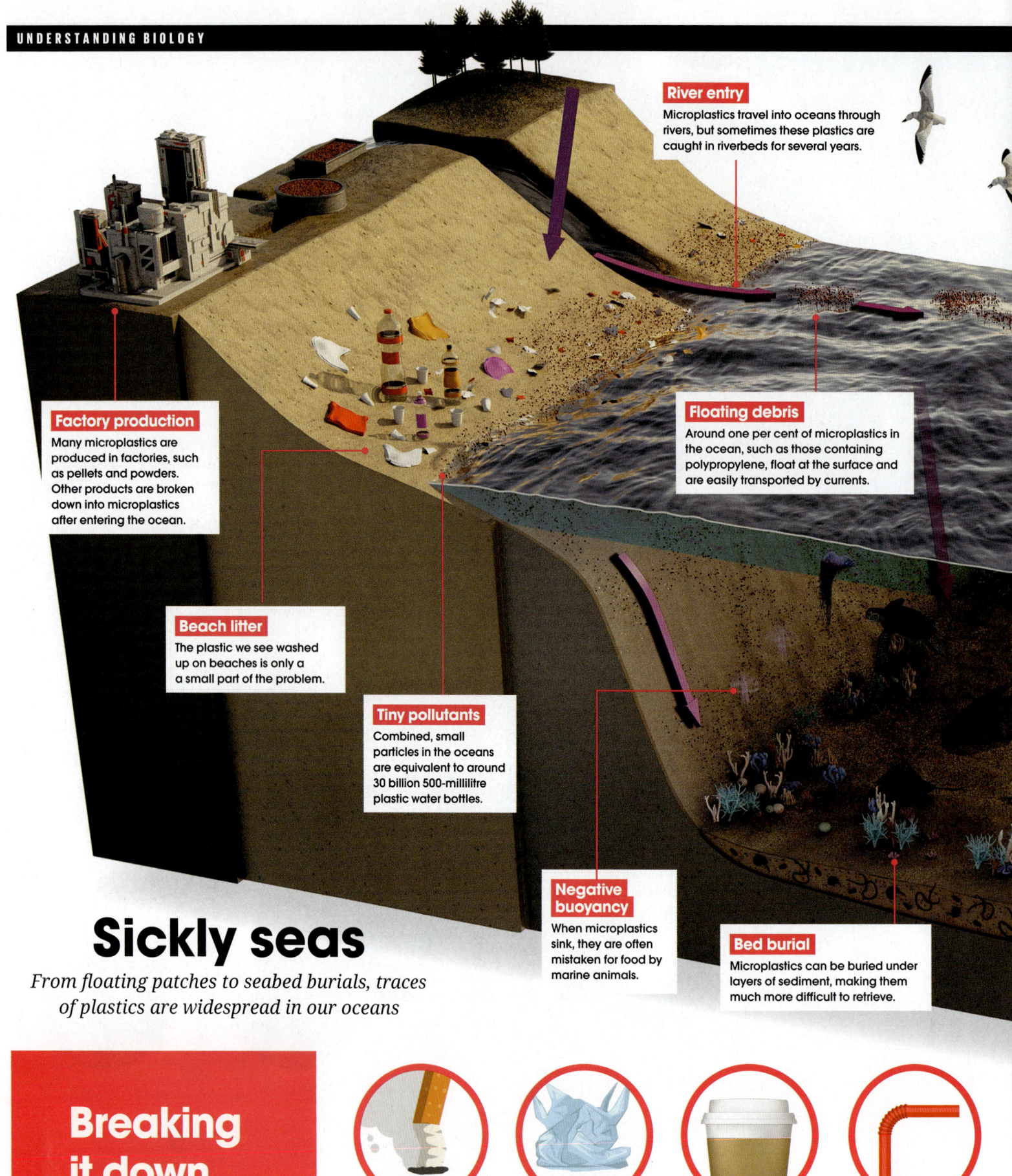

UNDERSTANDING BIOLOGY

River entry
Microplastics travel into oceans through rivers, but sometimes these plastics are caught in riverbeds for several years.

Floating debris
Around one per cent of microplastics in the ocean, such as those containing polypropylene, float at the surface and are easily transported by currents.

Factory production
Many microplastics are produced in factories, such as pellets and powders. Other products are broken down into microplastics after entering the ocean.

Beach litter
The plastic we see washed up on beaches is only a a small part of the problem.

Tiny pollutants
Combined, small particles in the oceans are equivalent to around 30 billion 500-millilitre plastic water bottles.

Negative buoyancy
When microplastics sink, they are often mistaken for food by marine animals.

Bed burial
Microplastics can be buried under layers of sediment, making them much more difficult to retrieve.

Sickly seas
From floating patches to seabed burials, traces of plastics are widespread in our oceans

Breaking it down
How long do common plastic items take to decompose?

5 Years — Cigarette butts

20 Years — Plastic bags

30 Years — Coffee cups

200 Years — Straws

MICROBIOLOGY

90

WHAT ARE MICROPLASTICS?

Did you know?
8.8 million tonnes of plastic entered the oceans in 2010 alone

Plastic gyres
Gyres form when ocean currents move in a circular pattern. These large systems, caused by global winds and Earth's rotation, accumulate large masses of microplastics.

Fishing contribution
Discarded fishing gear is one of the leading contributors to ocean microplastics.

Passing on plastic
Microplastics enter fish and other aquatic animals through their gills or mouths. As bigger animals eat their plastic-containing prey, these pieces accumulate up the food chain.

Zooplankton consumption
Small animals that are carried with the ocean current can consume microplastics, entering them into the first stage of many food chains.

What are ocean garbage patches

There are five main ocean gyres, or garbage patches, in Earth's oceans that have brought together many millions or billions of plastic pieces. In the largest of these, the North Pacific Gyre, there are an estimated 1.8 trillion pieces – that's about 250 pieces of plastic for every human on Earth. The other gyres are the North Atlantic Gyre, the South Atlantic Gyre, the South Pacific Gyre and the Indian Ocean Gyre. Although these patches are mostly made of plastic debris, they aren't exclusively plastic. Among the microplastic masses are larger pieces of rubbish, including fishing equipment, bottles and even mobile phones. Due to the circular movement of these patches, plastic items remain trapped in ocean gyres and are broken down by the Sun and the waves into microplastics.

The locations of the world's five main garbage patches are highlighted by the light circles

400 Years	**450 Years**	**500 Years**	**500 Years**	**500 Years**	**600 Years**
Can holders	Bottles	Toothbrushes	Nappies	Styrofoam	Fishing lines

UNDERSTANDING BIOLOGY

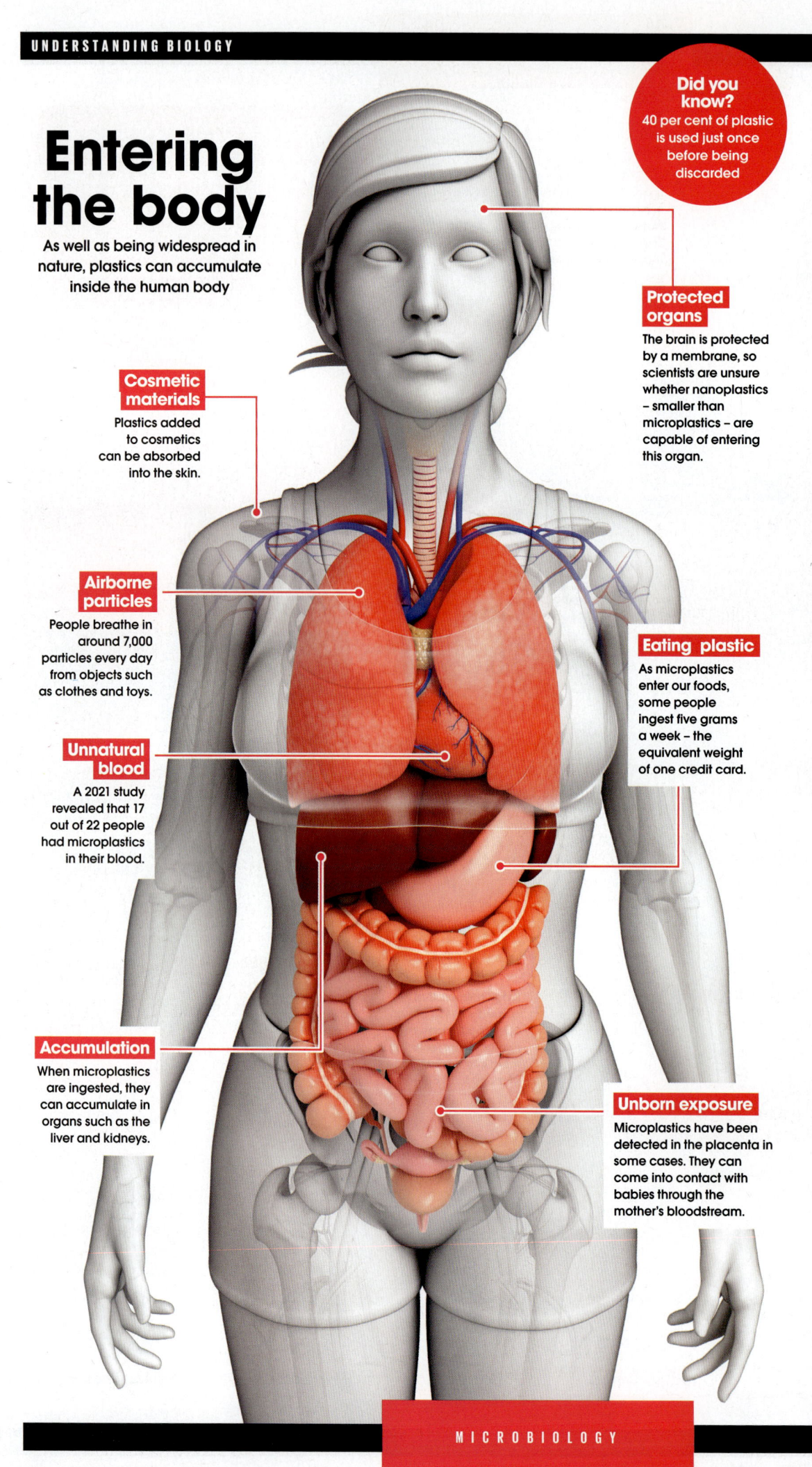

Entering the body

As well as being widespread in nature, plastics can accumulate inside the human body

Did you know? 40 per cent of plastic is used just once before being discarded

Protected organs
The brain is protected by a membrane, so scientists are unsure whether nanoplastics – smaller than microplastics – are capable of entering this organ.

Cosmetic materials
Plastics added to cosmetics can be absorbed into the skin.

Airborne particles
People breathe in around 7,000 particles every day from objects such as clothes and toys.

Eating plastic
As microplastics enter our foods, some people ingest five grams a week – the equivalent weight of one credit card.

Unnatural blood
A 2021 study revealed that 17 out of 22 people had microplastics in their blood.

Accumulation
When microplastics are ingested, they can accumulate in organs such as the liver and kidneys.

Unborn exposure
Microplastics have been detected in the placenta in some cases. They can come into contact with babies through the mother's bloodstream.

Reducing exposure

Microplastics appear to be everywhere on the planet. But is it possible to avoid exposure to them in everyday life? Because you can't always see microplastics, you don't know when they're entering your body, but there are steps you can take to reduce the volume you come into contact with.

Firstly, you can change the containers that you eat or drink out of. For example, replace disposable coffee cups with stainless steel or glass ones and stop microwaving foods in plastic containers. Plastic can enter your food much more easily when heated, so transferring it into another dish can reduce how much plastic you accidentally consume. Plastic tea bags are capable of releasing 11.6 billion microplastics and 3.1 billion nanoplastics into your drink. To avoid this, you can replace them with loose leaf tea. Similarly, when choosing meals to eat, cutting down on seafood consumption can reduce how many of these particles enter your body.

Every time you wash clothes made of nylon, polyester or acrylic, you release hundreds of thousands of microfibres. To avoid releasing these microplastics, you can check the materials of clothes before buying them or install a filter into the washing machine to catch any that come loose. Additionally, when choosing cosmetics to apply to your skin, it helps to choose products that don't contain any microbeads or any form of plastic ingredient.

Tea bags are often made using polypropylene to seal them

MICROBIOLOGY

WHAT ARE MICROPLASTICS?

Removing microplastics
These technologies and techniques can help reduce microplastics' environmental impact

Blue mussels filter their food from the water

Mussel power

In 2021, studies carried out at the UK's Plymouth Marine Laboratory found that mussels – and in particular their poop – could help scientists remove microplastics from rivers and estuaries. The experiment, which focused on blue mussels, uncovered that because these creatures can filter microplastic pieces out from their bodies and into their faeces, they become buoyant when excreted.

By using this method, a quarter of all waterborne microplastics that these mussels encounter can be released to the water's surface for collection. The scientists discovered that 300 blue mussels can filter out 250,000 microplastics per hour using this method. Since the experiment, trials have begun to find an effective system for mussel microplastic removal. These involve submerging masses of mussels in containers, with nets attached to collect the microplastic-filled faeces.

Robot removers

One potential solution for removing these tiny plastic pollutants is to deploy robots into the environment. In June 2022, a research paper from Sichuan University in China described a flexible, 13-millimetre-long fish-like robot that is both self-propelling and self-healing. The undisclosed soft material attracts the dyes and heavy metals in the microplastics to the robot fish through chemical bonds and electrostatics. Once in contact, the robot holds onto the microplastics and carries them until they are collected. If these fake fish get damaged while in the water, they can repair themselves and continue to depollute the seas, with a retained efficiency at least 89 per cent of what it was before it was damaged.

More research is needed before these robots can be released on a large scale to make a difference, but it's a useful insight into the technology that could one day be working in the background to reverse the damage we've done to the environment.

Preventing plastic water
How do wastewater treatment plants remove microplastics?

1 Sedimentation
Water moves slowly through a sedimentation tank to allow suspended microplastic particles to sink and be filtered out of the water.

2 Aeration
Oxygen is added to cause growth of microorganisms and break down organics. The mesh used to collect waste products can trap up to 99 per cent of microplastics.

3 Oxidation
An oxidant called ozone is added to water to break down contaminants such as microplastics. This is the most effective step to completely destroy them.

UNDERSTANDING BIOLOGY

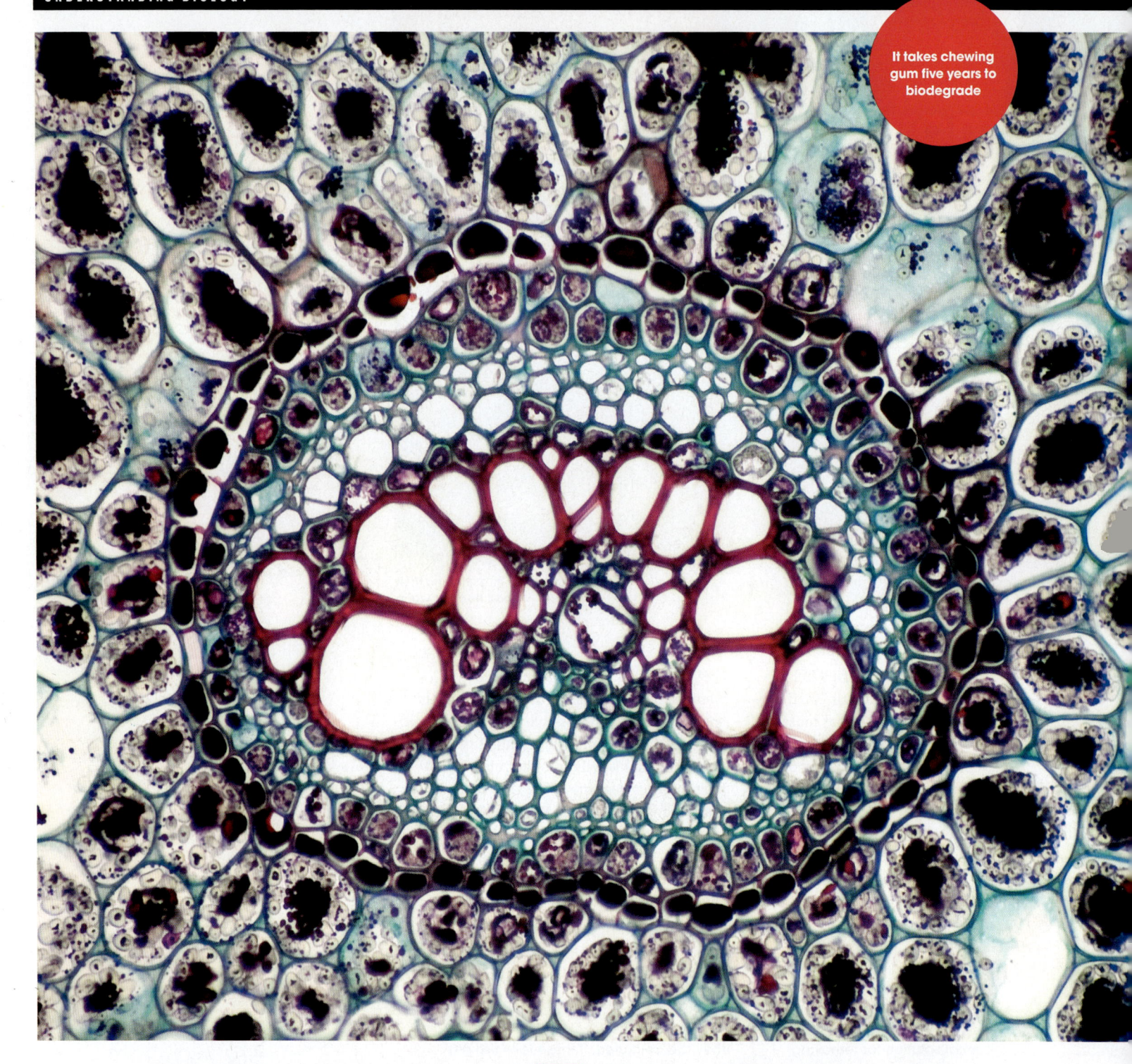

It takes chewing gum five years to biodegrade

What makes things biodegradable?

The chemistry and biology behind the natural breakdown of organic matter

MICROBIOLOGY

WHAT MAKES THINGS BIODEGRADABLE?

The term 'biodegradable' is often used to describe a material's ability to be broken down naturally by the environment through a process known as biodegradation. During this process, organic matter, such as that found in plants and animals, is torn apart, broken down and digested by fungi and microbes. What remains is a nutrient-rich biomass that fuels the growth of new plants and animals in a repeating and self-sustaining process – it's quite literally the circle of life.

One of the most important aspects of biodegradation is the transfer of carbon. Often referred to simply as the carbon cycle, this natural undertaking helps regulate the planet's temperature, as well as provide food and energy to its inhabitants. During the cycle, carbon is exchanged with oxygen in the atmosphere through plant photosynthesis, which is then stored in plant matter. That carbon is passed on to the animal that eats the plant, then to the animal that eats that animal, and so on, up through the food chain. When plants and animals die, that carbon is returned to the earth through biodegradation, whereby countless microorganisms chow down on organic matter, releasing carbon dioxide into the atmosphere to restart the cycle.

Outside of the natural order of things, the word 'biodegradable' is used as a label to describe products and packaging that are capable of undergoing the process of biodegradation, as opposed to non-biodegradable materials such as plastic, glass and metals. When we toss away our rubbish, it typically ends up in one of three places: an incinerator, landfill site or recycling centre. As the name suggests, incinerators torch waste and convert it into ash and gas, whereas recycling centres seek to transform materials into something useful. Landfill sites, on the other

A lack of oxygen inside a landfill site can make it difficult for material to decompose

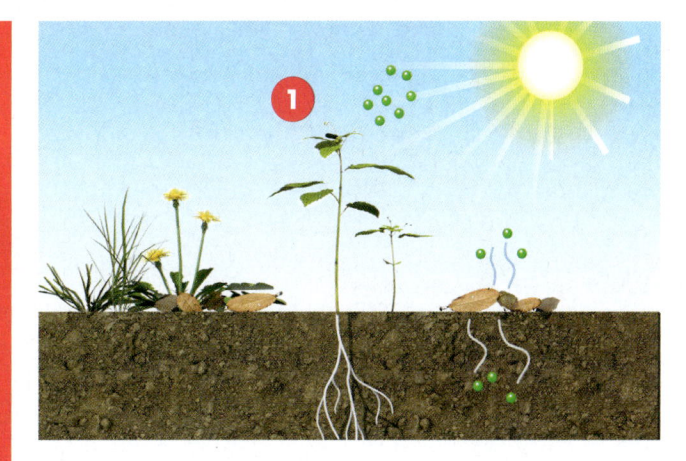

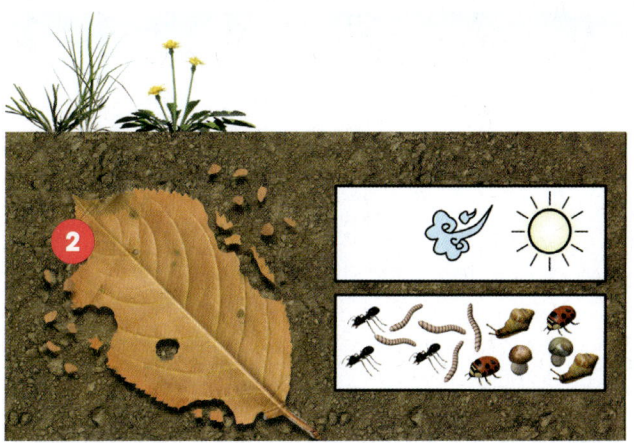

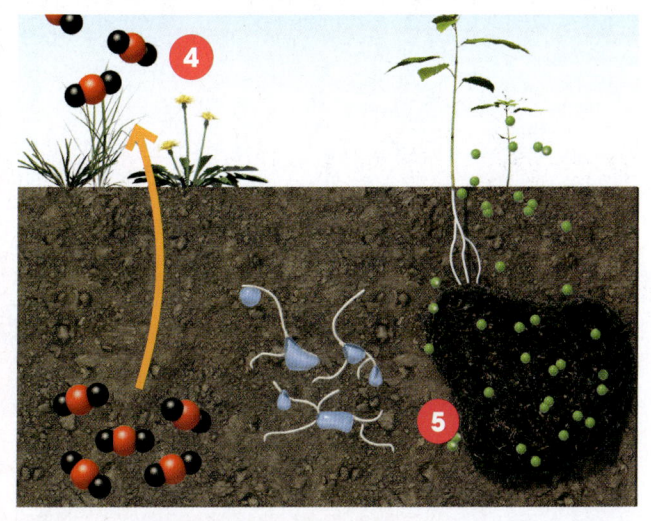

Cycling carbon
How biodegradation puts carbon back into the earth to release it again

1 Organic material
Organic matter is filled with carbon atoms that are recycled into the environment through biodegradation.

2 Decomposition
Organic matter decomposes into smaller pieces, either through physical forces such as wind and rain or from living organisms in the soil, such as fungi and invertebrates.

3 Digestion
Over time, microbes such as bacteria finish off and metabolise what remains of the organic matter.

4 Byproducts
Once organic matter has been completely broken down and devoured, what remains are three byproducts: carbon dioxide, water and a carbon-rich substance called biomass.

5 Assimilation
The byproducts of biodegradation are then assimilated back into the environment and used to promote the new growth of organic matter.

"One of the most important aspects of this process is the transfer of carbon"

UNDERSTANDING BIOLOGY

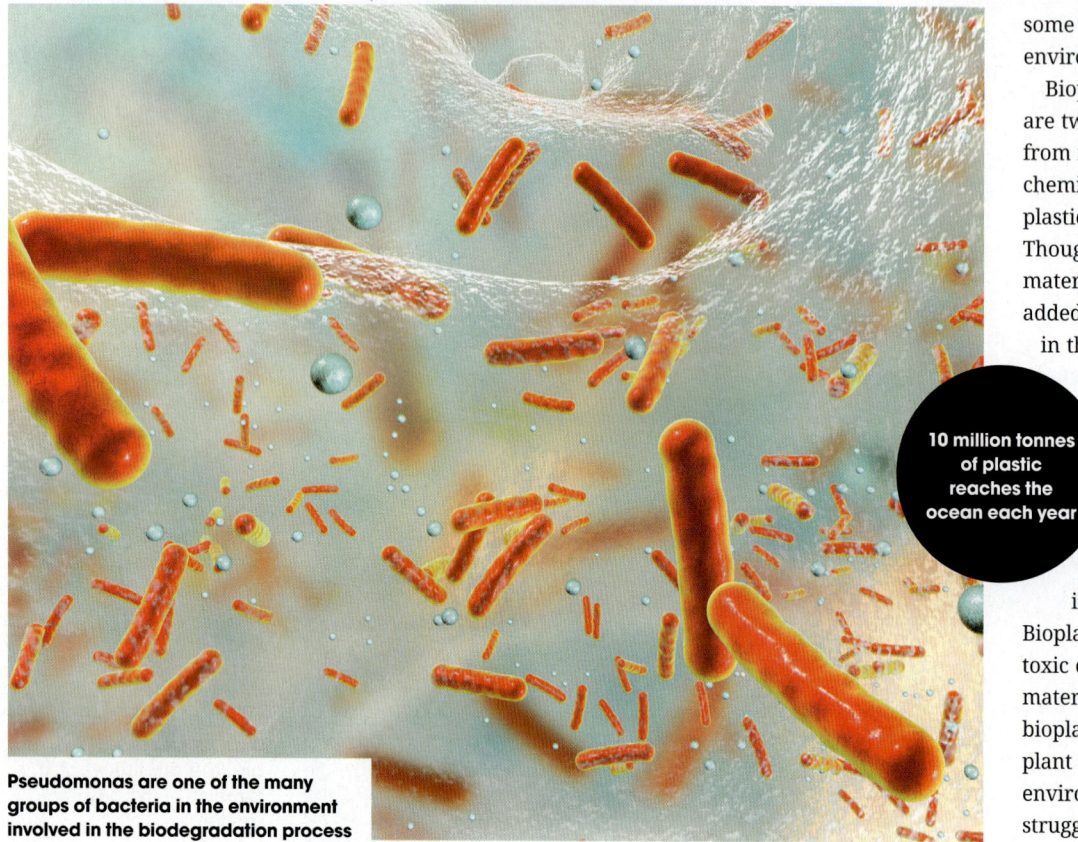

Pseudomonas are one of the many groups of bacteria in the environment involved in the biodegradation process

10 million tonnes of plastic reaches the ocean each year

hand, are places where biodegradation can occur. It occurs under one of two conditions: aerobically, with the help of oxygen, or anaerobically, without oxygen. Because of the compacted structure of a landfill site, biodegradation often occurs anaerobically, the slowest of the two conditions. Some studies have found that food items such as grapes and corn cobs are still recognisable 25 years after entering a landfill. Meanwhile, in an oxygen-rich compostable environment, it would take only weeks for them to decompose.

The phrase biodegradable is often used interchangeably with compostable, but the two are very different. For a product or material to be considered compostable there needs to be no harmful chemicals or substances that release during the process of biodegradation. There are some instances where materials such as plastics are classed as biodegradable, but in the process they also release toxic chemicals into the environment.

Each year we produce around 380 million tonnes of non-biodegradable plastics, and only 50 per cent of this is recycled. Traditional petroleum-based plastics are made from oils and gases that form robust molecular chains called polymers. These chains are so robust that they are unable to be broken down through biodegradation alone. However, some biodegradable plastics are more environmentally friendly.

Bioplastics and biodegradable plastics are two different things. Bioplastics are made from natural plant matter that's usually chemically treated to form strong polymer plastics called polylactic acid or polylactide. Though biodegradable plastics still use raw materials such as oil, different chemicals are added that allow the plastic to 'biodegrade' in the right circumstances, such as in high temperatures and under ultraviolet light. Although biodegradable plastics will break down much faster than the non-biodegradable versions, taking between three and six months, they release harmful chemicals and substances into the environment if they are not disposed of correctly. Bioplastics, on the other hand, release no such toxic chemicals. In the same way that organic materials shed carbon during decomposition, bioplastics also release carbon stored in the plant matter within them. Despite the risk to the environment, the popularity of bioplastics has struggled to match that of non-biodegradable alternatives. In 2022, only 1,142 tonnes of bioplastics were produced globally.

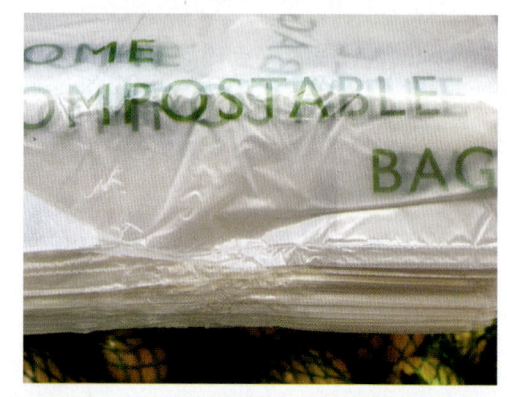

Bioplastics are often used in food packaging and bags

Biodegradable glass

Glass is typically made by heating natural raw materials such as sand or limestone. As a rigid non-biodegradable material, the only way glass can break down in the environment is through physical forces such as wind and water, which can take up to 4,000 years. In March 2023, researchers at the Chinese Academy of Sciences created an experimental glass made from modified amino acids and peptides. In the novel method, amino acids are heated and subjected to a supercooling treatment, then doused in water, rapidly forming a clear, glass-like material. When put to the biodegradability test, researchers discovered that the new glass material was broken down by microbes in soil within around three to seven months. Glass beads were also ingested by mouse subjects without causing harm and appeared to biodegrade in the body, suggesting there's scope for its use in drug delivery.

Advancements in bioglass could one day provide a new delivery system for medicines

MICROBIOLOGY

WHAT MAKES THINGS BIODEGRADABLE?

Growing plastic

How bioplastics are chemically engineered from crop plants

1 Plant materials
The building blocks for bioplastics come from natural sources such as corn, legumes, cassava and sugarcane.

2 Dissolution
Plant matter is broken down into starch, proteins and fibres using different acids.

3 Transforming starch
The starch is then separated from the rest of the solution, fermented and turned into lactic acid.

4 Linking chains
Starch is made up of lots of carbon chains, similar to those found in non-biodegradable plastics.

5 Forming process
The long molecule chains of lactic acid are injected into a mould and heated to set them into the desired shape.

6 Biodegradation
Microorganisms are able to break down bioplastics naturally, releasing carbon dioxide, water and biomass in the process.

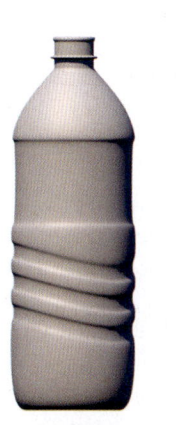

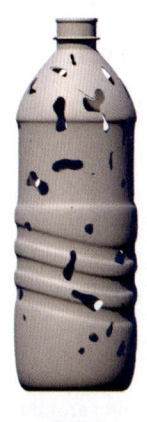

7 Going, going, gone
Bioplastics take around 12 weeks to break down.

Day 1 | Day 28 | Day 38 | Day 58 | Day 80

5 Materials that can degrade properly

1 Cardboard
Around 72 million tonnes of cardboard is produced each year worldwide. Cardboard is made from natural fibres that take around two months to break down.

2 Paper
Paper packaging takes around two to six weeks to decompose. A commonly recycled material, the cellulose fibres that make up paper can be recycled five to seven times before they become too weak to form paper.

3 Bamboo
Made from the fast-growing bamboo plant (Bambusa vulgaris), this type of biodegradable material can become compost in up to six months.

4 Organic fabric
Organic materials such as cotton or hemp biodegrade at different rates. Cottons may take several months to decompose, whereas hemp takes only a couple of weeks.

5 Cornstarch
As a replacement for polystyrene peanuts, cornstarch packing peanuts take around 90 days to decompose and even dissolve in water

UNDERSTANDING BIOLOGY

3 Cleanup
Algae grows quickly, and it efficiently removes harmful carbon dioxide, nitrates and phosphates from the air.

85 per cent of liquid biofuel that's produced globally uses ethanol

What is biofuel?

Biofuels are futuristic fuels that can power cars and trucks with crops, algae and even rubbish. Here's how they're produced

2 Growth Spurt
The sunlight causes photosynthesis, which is how plants turn carbon dioxide and water into oxygen and sugar.

Heat and maize are used to make bread, pizza and other delicious foods, but in a few years they could also power your car. Renewable biofuels could soon replace the harmful fossil fuels that we're used to using. It's now well-established that fossil fuels are incredibly harmful to our health and the environment, and they're not renewable, either. When so many people rely on petrol and diesel to power vehicles, it makes sense to develop a renewable alternative that's easy to use.

That's where biofuel comes in. The most common biofuel produced globally is ethanol, and it's used frequently in Brazil and the US, while biodiesel is more popular in Europe. Ethanol is a clear, flavourless alcohol that's created by fermenting and distilling sugary crops like wheat, corn and sugar cane. It's combined with petrol to make fuel more environmentally friendly, and it's already in widespread use. More than 95 per cent of petrol sold in the US contains ethanol, and E10 fuel – made from ten per cent ethanol – is now the standard petrol in the UK. Alcohol combines with oil or fat to create biodiesel. It works directly in many car engines, but it's usually combined with conventional diesel to create a more effective blend. It's produced from vegetable oil, animal fat, soy or palm oil.

There's a huge amount of potential in biofuel, and most of the big energy companies have already invested – but emerging energy sources like ethanol are first-generation biofuels, and they've got issues that need fixing before they can go mainstream. It currently takes more ethanol than gasoline to produce the same amount of energy, for instance. Production is expensive, and several parts of the process sometimes use fossil fuels, which means that some biofuels aren't actually carbon neutral.

Some environmental campaigners also say that it would be more useful to grow crops for food rather than biofuel, and that growing crops for biofuel can cause problems with soil erosion and deforestation. Using land for fuel rather than food can lead to an increase in food prices, too, and can hinder natural habitats. Crop and fat-based biofuels may not be perfect, but

Brazil and the US produce 87 per cent of the world's ethanol

MICROBIOLOGY

WHAT IS BIOFUEL?

The Algae Alternative

Algae is touted as a more efficient option for biofuel creation – but how does this work?

1 Specialist Facilities
Algae for fuel production is usually grown in a nutrient-rich water system called a bioreactor that's exposed to sunlight.

4 Water Mess
Once algae is harvested, it's usually dehydrated. After that, oil and carbohydrates can be used to create biodiesel or ethanol.

Extraction

5 Surface Pressure
Alternative methods involve subjecting algae to high-pressure and temperature environments to extract oil to use in biofuels.

Sweet Reactions

Creating ethanol is a complex chemical process. Plant cells contain cellulose, hemicellulose and lignin. Acids, enzymes and other chemicals break the plants down, and the result is a pure sugar solution called sucrose. At this point, scientists add yeast, and the solution heats to between 250 and 300 degrees Celsius. The yeast has an enzyme called invertase. Heat activates this, converting the sucrose into glucose and fructose. These sugars combine with another enzyme, called zymase, which converts them into ethanol. That ethanol still has lots of water, though, so the next step involves boiling. Because ethanol boils at 78.3 and water boils at 100 degrees Celsius, the ethanol boils and turns into vapour first; it can be separated from the water, condensed back to liquid form and used for biofuel.

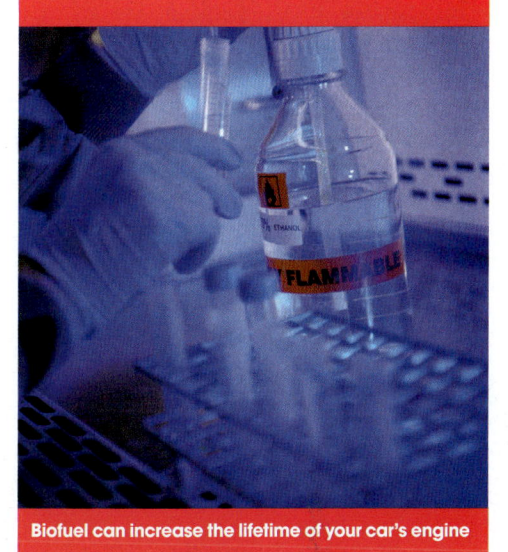

Biofuel can increase the lifetime of your car's engine

"Wheat and maize could power your car in a few years"

those aren't the only biofuels available – some organisations are creating biofuels with algae instead. This process uses water and land that often isn't suitable for many other situations, so it doesn't take up space that's useful for food production, and it often has better yields than other types of biofuel components.

There's plenty of development beyond algae, too. Many companies are developing crops for biofuels, and those will improve yields, increase efficiency and reduce costs. Some even use seaweed. Scientists are also working to extract biofuels from household waste, wood chips and other junk – a move that could massively increase the material that's viable for biofuel production. These second-generation biofuels could make biofuel far cheaper and more accessible, and help cut emissions down.

Racing Ahead

The World Rally Championship (WRC) is a global motorsport competition that made a groundbreaking switch to biofuel for its 2022 season. The WRC now uses a second-generation biofuel made from organic waste material, which is combined with synthetic fuel. The production process uses renewable energy and takes carbon dioxide out of the atmosphere thanks to carbon capture. It's the first time that a fossil-free fuel has powered a world motorsport championship. This isn't the only time you'll see biofuel used in a high-performance situation, either. Formula 1 now runs with a ten per cent biofuel blend, and the aviation and shipping industries are increasing their usage. It's even being used by space start-ups.

Top-tier motorsport competitions are now using biofuel to improve emissions and sustainability

99

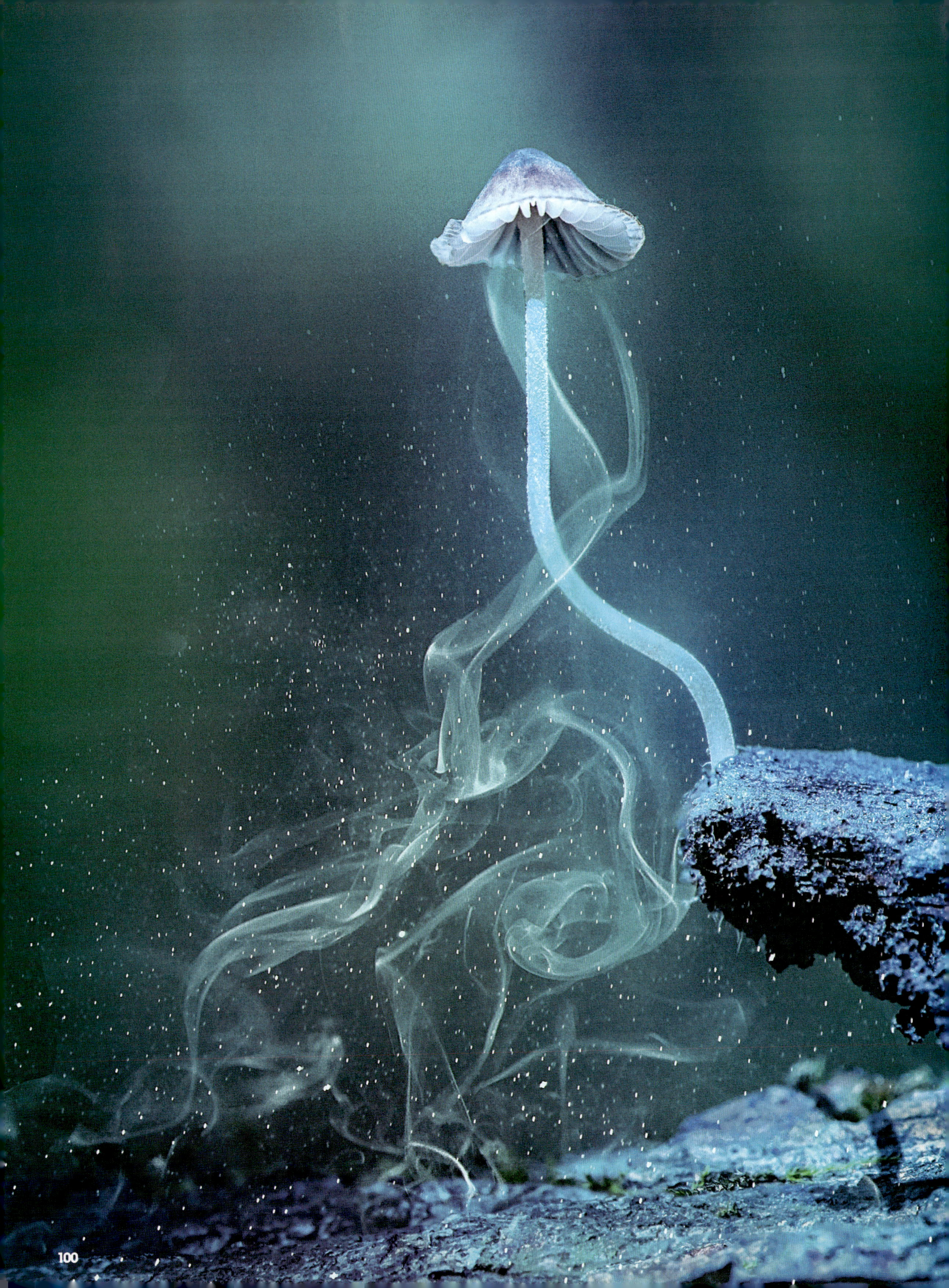

Plants & Fungi

UNDERSTANDING BIOLOGY

The oldest plants in the world

Discover the trees, flowers, shrubs and clonal colonies with deep roots in our planet's history

PLANTS & FUNGI

THE OLDEST PLANTS IN THE WORLD

Did you know?
Individual trees in the Huon pine colony can reach 3,000 years old

Huon pines are endemic to Tasmania

Mount Read Huon pine colony

10,500 years old • West Tasmania, Australia

The ancestry of the Huon pine plant family, Podocarpaceae, can be traced back to 200 million years ago through fossilised plant pollen records. But despite being called a pine, the Huon pine is actually another type of coniferous tree with scale-like leaves called a Podocarpus tree. In Tasmania, there's one clonal population of this tree that covers an area of 10,000 square metres. Even though it's a large population each one stemmed from a single male Huon pine tree that grew over 10,500 years ago. This makes it the world's oldest colony of genetically identical trees.

Today, each of these individual trees is an exact replica of the original, which shed its branches and so multiplied in number. After uncovering that each of the Mount Read Huon pine trees were identical in the 1990s, scientists developed a theory about how the original ancient tree lives on in others. The main theory is that its branches were weighed down by snow and ice over the course of millennia. As the branches came into contact with the ground they were able to re-root themselves and form new trees. While none of the trees today are as old as the first to plant its roots here, this cloned population is essentially the same organism as the one that shed its branches centuries ago.

King Clone

11,700 years old • Mojave Desert, California

The survival abilities of creosote bushes are impressive across the board; the small, waxy-leaved shrubs are adapted to life in the most arid deserts. However, the doughnut-shaped bushes that make up the King Clone bush ring are believed to have come from one of the first life forms to thrive in the Mojave Desert, appearing shortly after glaciers covered this land. When the original plant that grew in place of this ring began to wither away thousands of years ago, its roots branched out underground and created offshoots from which new plants of the same organism could grow. Over time, these plants have expanded to form the ancient bush ring, with a diameter that reaches up to 20 metres and has an average of 14 metres.

University of California professor Frank Vasek discovered that King Clone was a single organism and estimated its age

UNDERSTANDING BIOLOGY

Did you know?
The Hundred-Horse Chestnut is now a national monument

Ancient individuals
Where can you find the oldest trees in the world?

1 Methuselah
4,854 years old
In the White Mountains of California, a chunky bristlecone tree reaches to the sky like a twisted claw. The bends of each branch are the final remnants of the tree, which germinated before the Egyptian pyramids were built. Only a small portion remains living, and continues to bear pinecones and needles. It measures 70 metres tall and has a diameter over eight metres at ground level.

2 Sarv-e Abarkuh
4,500 years old
This bushy evergreen can be found at the heart of Yazd, Iran, in the ancient city of Abarkuh. The Persian cypress tree symbolises long life and beauty in Iranian culture and is likely to be the oldest living organism on the Asian continent.

3 Llangernyw Yew
4,000 years old
Since the Bronze Age, this prehistoric yew tree has been growing in the rural village of Llangernyw in North Wales. The exact age of the tree is difficult to determine due to the original core trunk giving rise to newer wood. When the Celts settled in Llangernyw, they considered the tree to be sacred as it could revive itself with new bark.

4 Alerce Milenario
5,480 years old
In Alerce Costero National Park, Chile, a conifer tree over 60 metres tall and with a diameter of 4.26 metres extends from a damp ravine. This ravine may be the secret to its long life, with the steep walls sheltering the tree from the elements, such as storms and fires. Scientists believe that this tree began to grow at the same time that humans started writing.

5 Patriarca da Floresta
3,000 years old
This tropical tree, of the species Cariniana legalis, is a semi-deciduous tree with an umbrella-shaped top. Its name translates to 'father of the forest', and the tree is currently being protected by conservationists from the mass-scale deforestation that's occurring in Brazil.

PLANTS & FUNGI

THE OLDEST PLANTS IN THE WORLD

6 Olive tree of Vouves
3,000 years old

This majestic olive tree appears to hover over the ground due to its exposed hollow trunk. Located in Crete, Greece, this is the world's oldest olive tree, yet it still bears fruit today. Its wood is constantly renewing outwards, forming a 4.6-metre-wide trunk for the mere 6.5-metre-tall plant.

7 Jōmon Sugi
2,170 to 7,200 years old

On the Japanese island of Yakushima is a Japanese cedar tree older than Japan itself. Material called tephra, which is produced by a volcanic eruption, has been found in the layers of the tree and is estimated to have originated from the Yakushima volcanic eruption 7,300 years ago.

8 Hundred-horse chestnut
3,000 years old

If you were going to plant a long-lived tree, you might want to avoid placing it near an active volcano. However, Mount Etna in Sicily, Italy, is home to the world's largest and oldest chestnut tree. The tree has a diameter of 22 metres and is rooted less than eight kilometres from the volcano's crater.

9 General Sherman
2,200 years old

At 83 metres tall and with an 11-metre base diameter, General Sherman is the largest tree in the world by volume. The giant sequoia is located in California and has remained stable over thousands of years due to its widespread roots that hold onto the roots of other trees.

10 Old Tjikko
9,500 years old

This is the world's oldest tree, but the section that lies above the ground is only a few hundred years old and five metres tall. The spruce tree of Sweden's Fulufjället Mountain grew from the root shoots of its previous bark, which began growing just after the last ice age.

UNDERSTANDING BIOLOGY

King's lomatia
43,600 years old • Tasmania, Australia

There are around 300 of these plants left in existence on the planet, but some would say there's just one left. This is because the King's lomatia is a clonal group with the same genetic information. These flowering plants don't produce seeds, but shed their branches in order to reproduce identical copies through re-rooting.

The last remaining colony exists in a secret location in Tasmania in order to keep them protected. The plants of this colony have been growing in Tasmania for at least 43,600 years, but some scientists think it could be as many as 135,000. Today the King's lomatia covers a single one-kilometre strip of Tasmanian land.

Silene stenophylla has white blossoms and small seeds

Ice age flower
32,000 years old • Kolyma River, Russia

Some of the world's oldest plants have survived across millennia. But Silene stenophylla, a flowering plant known commonly as the narrow-leafed campion, has the honour of being one of the world's most ancient living plants after being brought back to life. When scientists were studying the remains of an ancient squirrel burrow, inside they discovered the seeds of the narrow-leafed campion, preserved in permafrost. Permafrost is any layer of soil that remains frozen for at least two years. In this case, however, the plant's genes stayed frozen for many thousands of years at over 40 metres below Earth's surface. By germinating these seeds, the scientists brought the preserved genes to life and were able to analyse the evolution of the species with this live, historic sample.

King's lomatia was discovered in 1937

Losh Run box huckleberry
13,000 years old • New Bloomfield, Pennsylvania

While covering an extensive area of Pennsylvania's forest floor, the Losh Run box huckleberry only reaches around 30 centimetres in height. The plant produces both bell-shaped flowers and dark berries that are similar to blueberries in appearance. It's the oldest known shrub colony in the world and was discovered in 1920 in Perry County, Pennsylvania. Losh Run remained a single colony until the 1970s, when a road was built to run through the area. Despite being separated, some of the oldest shrubs still remain dispersed around the area.

Box huckleberries bloom in May and June

PLANTS & FUNGI

THE OLDEST PLANTS IN THE WORLD

Pando, the oldest living organism

80,000 years old • Fishlake National Forest, Utah

Comprising over 40,000 quaking aspen trees, Pando is the world's largest organism by weight. It's considered a single organism because Pando, which translates to 'I spread' in Latin, is one large root system. The forest of trees covers 429,000 square metres and has an estimated weight of 6,600 tonnes. To make new individuals in the colony, Pando releases new shoots from the existing tree roots. These eventually separate from the original tree and grow a replica, increasing the chances of the organism's survival. This technique has allowed Pando to thrive over many thousands of years, combating forest fires and other natural disasters.

Did you know? Pando lost its title of 'world's largest organism' to a fungus

The 47,000 trees of Pando are considered the stems of a single organism

Welwitschia: The living fossil

This hardy 2,000-year-old plant is thought to have evolved with the dinosaurs

Namib-Naukluft Park, Namibia

Did you know? Each Welwitschia plant only produces two leaves

4 Leaves
Welwitschia leaves are adapted to absorb the limited water from the fog that forms in the desert during the night.

3 Dead apex
The top of the Welwitschia stem dies, preventing upward growth and causing the leaves to grow along the ground.

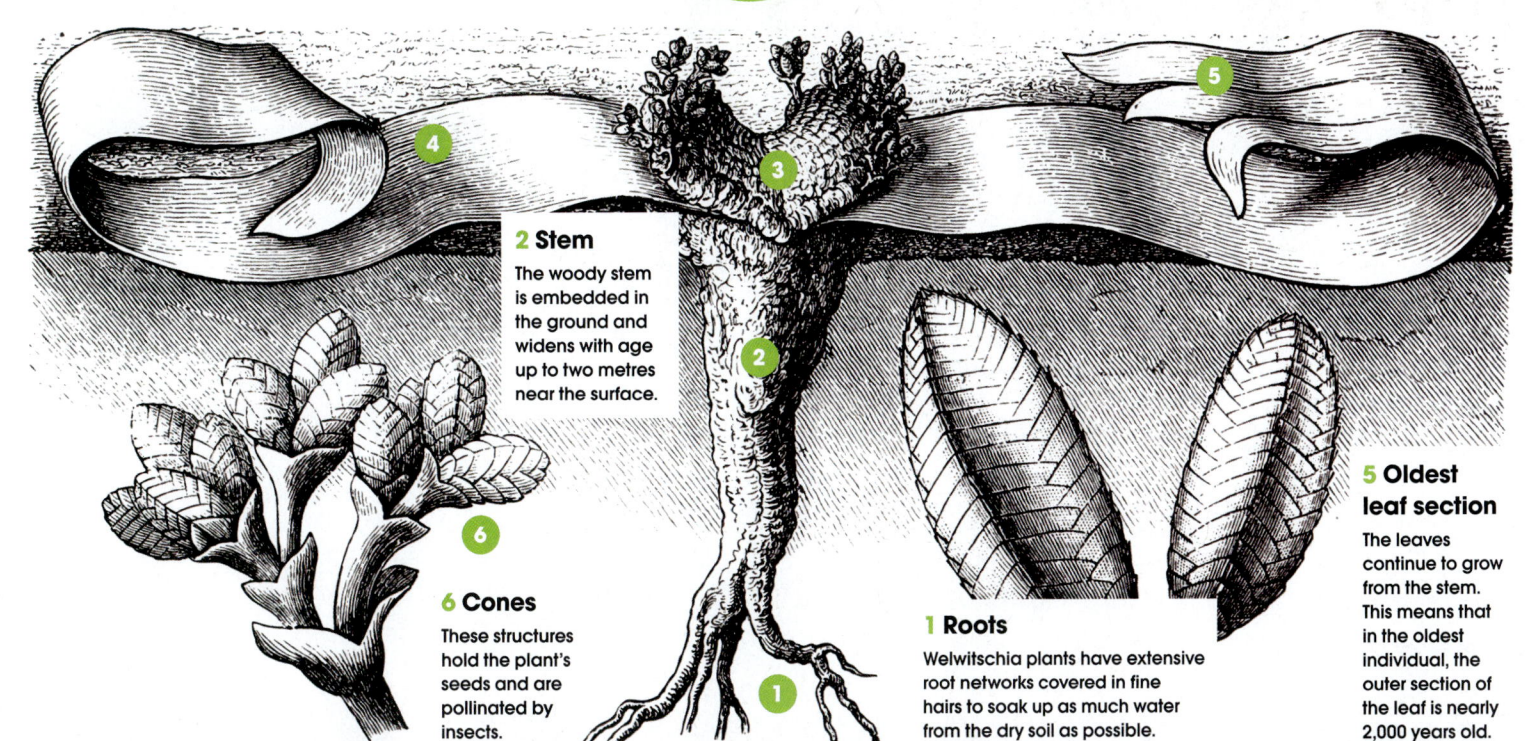

2 Stem
The woody stem is embedded in the ground and widens with age up to two metres near the surface.

6 Cones
These structures hold the plant's seeds and are pollinated by insects.

1 Roots
Welwitschia plants have extensive root networks covered in fine hairs to soak up as much water from the dry soil as possible.

5 Oldest leaf section
The leaves continue to grow from the stem. This means that in the oldest individual, the outer section of the leaf is nearly 2,000 years old.

UNDERSTANDING BIOLOGY

PLANTS & FUNGI

Fascinating fruit

From the stench of the durian to the flavour-changing ability of miracle berries, discover the wonderful world of fruit

The world is full of interesting and delicious fruits. From sour grapes to giant pumpkins, there are around 2,000 different types of fruits growing around the globe. In botanical terms, a fruit is a ripened ovary, a reproductive structure that bears the plant's ovule, also known as the seeds. Once a plant is pollinated, fruits begin to emerge from the fertilised flower and develop, mature and ripen, by which time the encapsulated seeds are ready for dispersal.

When it comes to identifying fruits, there are a whole host of subcategories they fall into. However, they largely sit within one of three categories: simple, aggregate and multiple. Simple fruits include the majority of fruits, such as stone fruits, pome and berries. Their anatomy is typically divided into three parts; the exocarp, mesocarp and endocarp. The outer skin is its exocarp, the mesocarp forms its flesh and the endocarp forms the innermost part of the fruit – the seeds.

Aggregate fruits, on the other hand, are merged from individual flowers that give rise to a curious collection. Raspberries are composed of lots of once-separate carpels – the juicy, seed-containing parts. Then there's the third kind of fruit, which pineapples belong to, called multiple fruits. Unlike aggregate fruits, multiple fruits are formed from a group of flowers that each produce a fruit that matures into a single mass.

Plants are proficient in the art of seduction. In lieu of attracting another plant for reproduction, fruits engage the senses of the animals needed for pollination and to spread their seeds. Seed dispersal by way of animal intervention typically occurs in one of two ways. When an animal such as a deer wanders into an orchard and chews down on juicy apples, it feeds not only on the flesh, but the encased seeds as well. From there, the seeds travel through the animal's digestive system and emerge once again among a pile of freshly made droppings. This nutrient-rich manure then provides the conditions needed for the seeds to germinate and a new plant to grow. Other animals, such as many rodent species, bury fruits in an attempt to store them for a later meal. However, when they leave the fruit beneath the soil, the seeds are given the opportunity to germinate and grow.

In the same way the vibrancy of flower petals attracts pollinators, fruits come in all colours of the rainbow in the hope of attracting potential seed-dispersing animals. Studies have shown that mammals such as monkeys and apes disperse more seeds of green fruits, while birds prefer the vibrancy of red fruits. Similarly, smell and taste also act as a way to entice animals to take a bite and spread the seeds within. In Madagascar, several species of colour-blind lemurs play an important role in spreading the seeds of fig trees. As colour isn't an alluring option for these trees, they have evolved fruits that emit a pungent scent when ripe, which the lemurs find enticing.

While many people will be able to empathise with the lemurs' love of sweet-smelling figs, not all fruits boast such appealing aromas. Pungent odours, including the smells of rotting onions and sewage, are equally important in attracting animals. For example, the noni fruit that's found on Pacific islands and in Southeast Asia, gives off an extremely pungent smell of stinky cheese and rotting fish. However, several bat and bird species aren't at all deterred by the terrible smell, instead flocking to this foul-smelling feast and subsequently spreading the seeds far and wide.

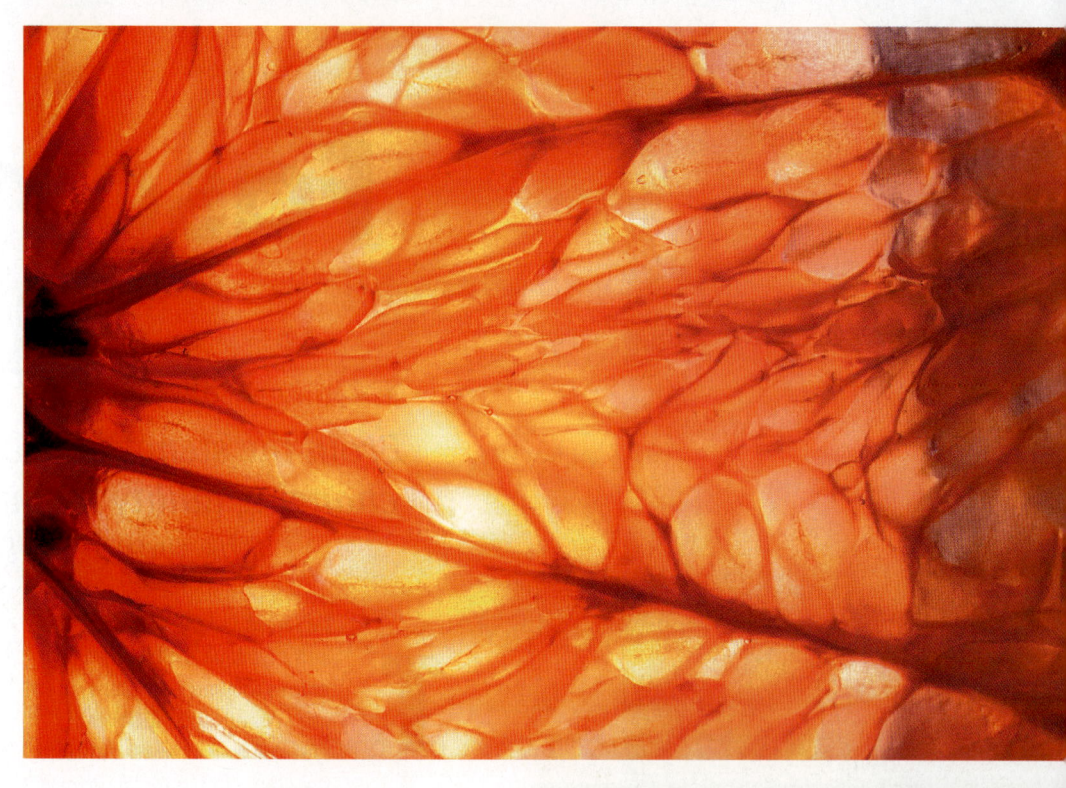

UNDERSTANDING BIOLOGY

Animals typically access a plant's fruit from the swinging branches of trees and shrubs, but some fruits have begun growing underground. Researchers discovered in 2023 that a new species of palm called Pinanga subterranea is growing its fruits and flowers below the surface. Around 171 species of plants are known to grow their fruits underground, including several orchids. Unlike topside fruits, those that are produced by the palm are not exposed to the breadth of pollinators and animals involved in seed dispersal. Scientists believe that subterranean insects may be responsible for pollinating these palms, and wild boar may sniff out and dig up the fruit for seed dispersal.

As well as being a vessel for plant propagation, fruits come with a whole host of health benefits for humans to enjoy. From topping up the body's vitamin levels to producing vital fibre needed for a healthy gut, the world's edible fruits are a great source of nutrition, with equally great health benefits. Many fruits such as blueberries, apples and cherries are also high in molecules called antioxidants. Once ingested, antioxidants seek out and destroy harmful molecules called free radicals. Free radicals are naturally produced by the body, as well as exposure to pollution and cigarette smoke, but can contribute to several health conditions, such as the buildup of cholesterol plaque, which leads to the development of heart disease. As well as cardiovascular issues, free radicals have been linked to chronic health problems such as inflammatory diseases and even cancer. When some fruit-sourced antioxidants come into contact with a free radical, they knock off one of its electrons and render it useless, preventing it from causing further harm to the body. To reap all the health benefits that fruits have to offer, the UK's National Health Service recommends eating five 80-gram portions of fruit a day for adults.

> **"Fruits come with a whole host of health benefits for humans to enjoy"**

Polejaeva studies the production of transgenic cloned animals for use in agriculture and biomedicine

Finger lime

Finger limes have bubbles of juice on the inside

Unlike supermarket limes, the Australian-native finger lime (Citrus australasica) erupts with limey caviar when it's squeezed. Found hanging from small finger lime trees in New South Wales, this cylindrical fruit might easily be mistaken for a young cucumber. Inside the fruit are individual sacs of lime juice that surround the central seeds of the plant. Like bubble wrap, the juice sacs help protect the seed pulp when the fruit falls from its tree. The sacs of juice provide a refreshing burst of sourness when you bite into them.

PLANTS & FUNGI

FASCINATING FRUIT

The miracle fruit
Tasteless alone, these berries have a unique property

Miracle berries (Synsepalum dulcificum) are native to the tropical regions of West Africa but have been cultivated around the world for their miraculous abilities. Despite being tasteless, miracle berries possess a protein that transforms the taste of other acidic or bitter foods. After chowing down on these berries, they will make taking a bite out of a lemon taste like you're drinking the sweetest lemonade and make a swig of apple cider vinegar bearable.

The mysterious powers of the fruit come from a natural sweetener called miraculin. When ingested, miraculin binds to the sweet taste receptors on the tongue for up to around 60 minutes. When a bitter-tasting food comes into contact with the miraculin, it changes shape, causing the sweet receptor to think it's come across a molecule from a sweet-tasting substance such as sucralose or aspartame. Even though miracle berries might make a great sugar or artificial sweetener substitute, the fruits are only produced on a small scale and are not able to replace what's currently commercially available.

Miracle berries grow on an evergreen shrub native to West Africa

Horned melon
Kiwano (Cucumis metuliferus), also known as the African horned cucumber or horned melon, is an intimidating fruit that originated in the Kalahari Desert, Africa, but has been cultivated in the tropical regions of South Africa and other parts of the world, such as New Zealand. The contrasting colours of the fruit's outer orange skin and its vibrant green pulp entice a myriad of critters, including birds, primates, rodents and antelopes, to feast and spread its seeds. The pulp is said to taste like a cross between a banana, melon and lime, and is packed with vitamins B and C.

A sweet miracle
How miraculin deceives your taste buds

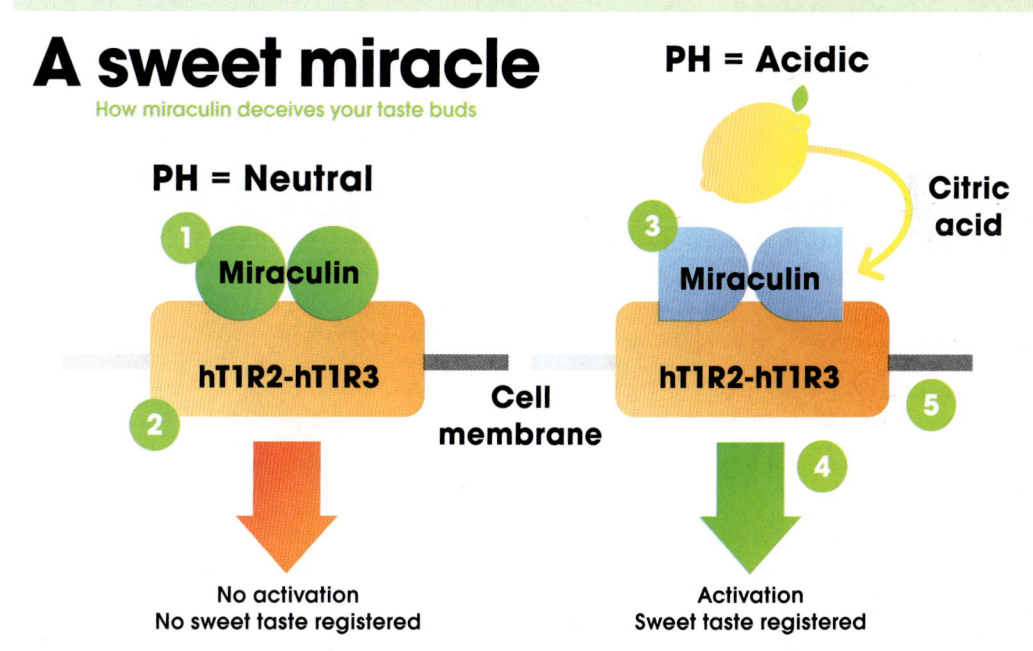

PH = Neutral
1. Miraculin
2. hT1R2-hT1R3
Cell membrane
No activation
No sweet taste registered

PH = Acidic
3. Miraculin — Citric acid
hT1R2-hT1R3
4.
5.
Activation
Sweet taste registered

1 Inactivated miraculin
Once a miracle berry is ingested, the miraculin binds to the sweet taste receptors and remains inactive until triggered.

2 Sweet receptors
Miraculin binds to a particular receptor called hT1R2-hT1R3.

3 Acid activation
With acids, such as the citric acid within a lemon, the miraculin changes shape and activates the sweet taste receptor.

4 A sweet taste
The sweet receptor that's bound to the miraculin registers a sweet taste, rather than an acidic one.

5 Cell membrane
Sweet taste receptors can be embedded in the cell membranes of the taste buds, called papillae.

UNDERSTANDING BIOLOGY

5 Facts: Fruits that smell

1. Jackfruit (Artocarpus heterophyllus)
Found in tropical and subtropical parts of the world, the jackfruit has a rotten-onion scent, but with a pleasant hint of pineapple and banana.

2. Cempedak (Artocarpus integer)
As with many fruits, the smell they emit intensifies the riper they get. For cempedak fruit, an already-potent aroma turns to the smell of ammonia when overripe.

3. Wood-apple (Limonia acidissima)
In the forests of Southern India you might find the interesting aroma of wood-apples, which has been described as a combination of fruity raisins and blue cheese.

4. Marang (Artocarpus odoratissimus)
The presence of a marang fruit will quickly fill a room with an odour reminiscent of gasoline, despite having a fruity pear-and-pineapple taste.

5. Pedalai (Artocarpus sericicarpus)
Found in Borneo and the Philippines, this melt-in-your-mouth fruit grows with an unusual hair-like husk and bears a scent resembling meringue.

Snake fruit

As its name suggests, snake fruit (Salacca zalacca) has a unique surface, akin to that of king cobra scales. The unusual fruit is grown in Southeast Asia and is native to Indonesia. Unlike many other drupes, the snake fruit has a leathery peel that's made up of lots of small, triangular scales. When the scales are chipped away, three lobes of juicy, white flesh are revealed. Snake fruit tastes like a combination of apple and pineapple, with the crisp texture of an apple, and is often eaten raw or turned into jams or syrups.

Did you know? Humans and bananas share 50 per cent of their DNA

The Southeast Asian rambutan is a relative of the lychee

Buddha's hand citron

A popular fruit within Asia, Buddha's hand citron (Citrus medica digitata) is thought to have reached China in the satchels of Buddhist monks travelling from India in the 4th century. Other than its digit-like appearance, one of the remarkable things about this fruit is its high level of pith. Citruses, such as lemons and limes, all have an internal fibrous pith to support juicy seed-bearing centres. However, the majority of this fruit is made up of pith, which limits its use. Traditionally, Buddha's hand is used to enhance the scents of other foods, such as sugars and ice cream, thanks to its potent tangerine-and-lemon aroma.

The fruit is sometimes used as a religious offering

PLANTS & FUNGI

FASCINATING FRUIT

The king of all fruits
The overwhelming aura of durian

Rotting onions, old socks, bin juice and raw sewage are all ways in which the smell of the durian fruit (Durio zibethinus) has been described. The smell of durian is so potent that consumption of the fruit has been banned on public transport in Singapore, and in some hotels in Thailand. A durian's scent stems from its ripened pulp and is the result of the release of a combination of more than 50 compounds, including an odorous sulphur-smelling chemical called ethanethiol. On their own the individual compounds aren't particularly powerful in their odours, nor does any smell like a durian. However, when they all come together they create the symphony of durian smell. The reason behind the odorous nature of this fruit is like any other: to entice animals that find the scent appealing in the hope of dispersing the seeds within.

Despite its pungent aroma, durian, also nicknamed the 'king of fruits', is an incredibly popular food, particularly in many Southeast Asian countries. As a nutritious fruit, durian is packed with helpful vitamins and minerals that support the body's immune system. Studies have also shown that the fruit is full of folate and folic acid, which can help combat anaemia and promote regular tissue growth during pregnancy. From coffee to candy, the distinctive scent and taste of durian fruit has also found its way into countless snacks and confectionery around the world.

Although it might smell strong, ripe durian has a sweet, caramel taste and a custardy texture

Durian anatomy
Where does its infamous odour come from?

Peduncle
The stalk of the durian supports the flower the fruit grows from.

Spines
The spines that coat the husk of the durian not only deter unwanted predators, but also reduce the impact on the flesh inside when the fruit falls.

Aril
Also known as the pulp, this edible part of the durian is the source of the majority of its potent aroma.

Husk
The external skin of the durian is covered with spines to prevent animals from prematurely taking a bite.

Locules
Within the fruit are segments or chambers where the pulp and seeds can grow.

Abscission zone
When fruit is ripe and ready to drop to the ground, the plant cells between the fruit and parent plant separate at a junction called the abscission zone.

Chocolate pudding
While hanging from its parent plant, the black sapote (Diospyros nigra), also known as the chocolate pudding fruit, resembles an unripened tomato. However, give it time to ripen and it transforms into a delicious dessert. After ten days of picking the chocolate pudding fruit and storing it at room temperature, the flesh within will turn a rich, gooey brown and gain a sweet, pudding-like flavour, with a hint of banana. Black sapote is full of nutrients, including around four times the amount of vitamin C found in an orange.

Black sapote tastes like a dessert when it's ripe

UNDERSTANDING BIOLOGY

Why leaves turn brown in autumn

Earth's deciduous plants lose their luscious green leaves during this changing of the seasons

In many parts of the world, the end of the summer season is marked by drastic changes in the trees' appearance. Leaves of red, orange, yellow and brown dance through the breeze before landing to create a crunchy layer on the ground. Autumn can produce distinctive colours, but for some plants it's a time to take drastic action for survival.

Evergreens aside, trees have very thin leaves. To keep them attached throughout winter would be dangerous as they would freeze – their cells would rupture and lose vital nutrients. The usual food-making process of converting sunlight and carbon dioxide into sugars is put on hold. The production of green cells called chloroplasts that enable this process is halted and nutrients are absorbed into the core of the tree to be utilised during the colder months. With this reduction of the core green cells, other pigments that exist in the leaves are given a season to paint the landscape with their bold colours.

On the ground, a leaf will decay as microbes in the soil help it decompose

What triggers new colours?

One of the most important factors in triggering these drastic changes is the amount of daylight available. As the days grow shorter and shorter, plants are unable to rely primarily on sunlight for their energy production. Secondly, the rate of photosynthesis decreases as the temperature plummets. The combination of these seasonal changes means that when autumn sets in, plants can't thrive using the same systems.

When the tree has absorbed the useful nutrients from a leaf, it detaches from the tree

PLANTS & FUNGI

WHY LEAVES TURN BROWN IN AUTUMN

Autumn anatomy

The cells and structures that help leaves change colour

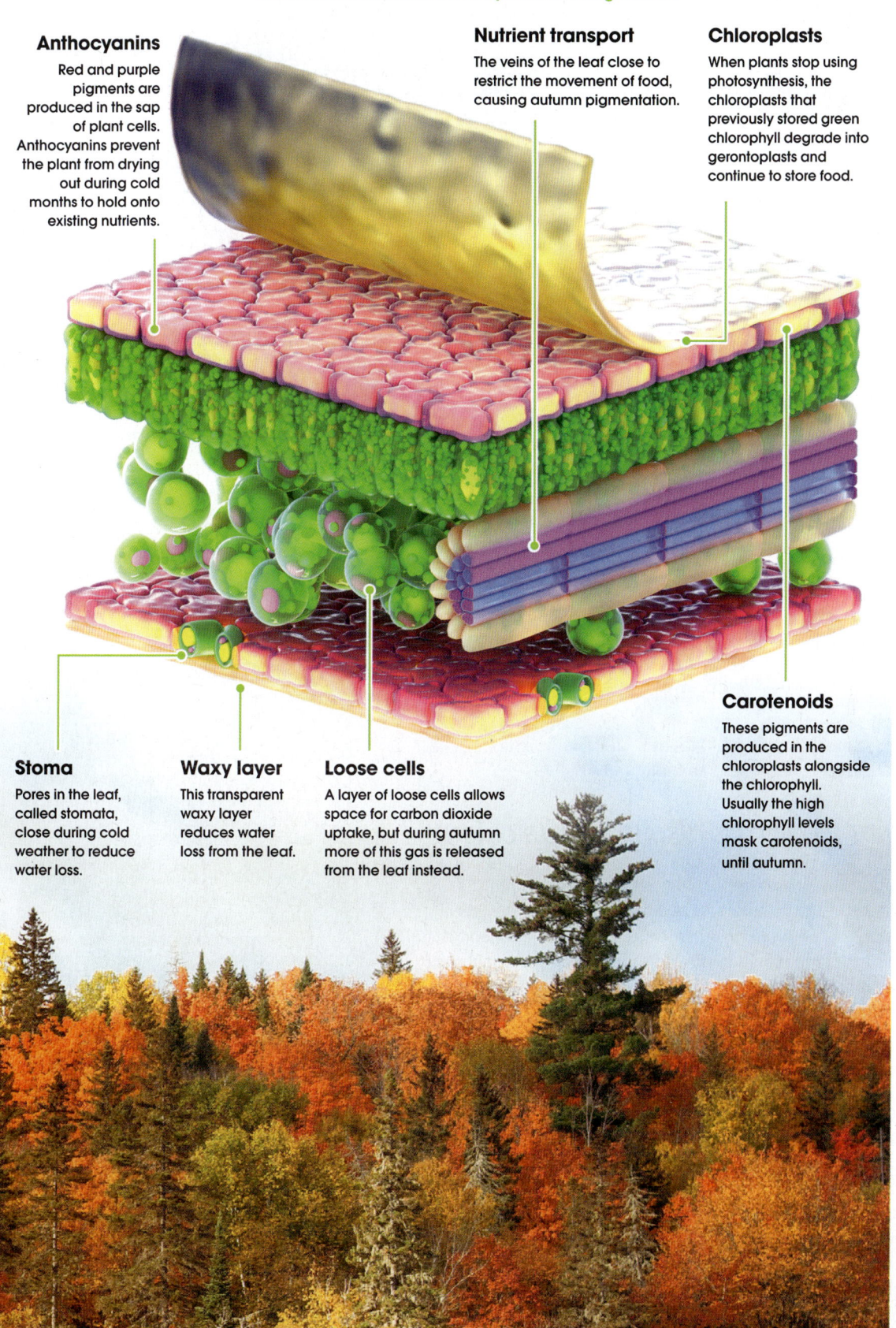

Anthocyanins
Red and purple pigments are produced in the sap of plant cells. Anthocyanins prevent the plant from drying out during cold months to hold onto existing nutrients.

Nutrient transport
The veins of the leaf close to restrict the movement of food, causing autumn pigmentation.

Chloroplasts
When plants stop using photosynthesis, the chloroplasts that previously stored green chlorophyll degrade into gerontoplasts and continue to store food.

Stoma
Pores in the leaf, called stomata, close during cold weather to reduce water loss.

Waxy layer
This transparent waxy layer reduces water loss from the leaf.

Loose cells
A layer of loose cells allows space for carbon dioxide uptake, but during autumn more of this gas is released from the leaf instead.

Carotenoids
These pigments are produced in the chloroplasts alongside the chlorophyll. Usually the high chlorophyll levels mask carotenoids, until autumn.

5 Colours in the autumn palette

1 Brown
Compound: Tannins
These compounds are waste products of metabolic processes within the leaves. While they are always present, only when other pigment levels, such as chlorophyll and carotenoids, are low are these visible as brown hues.

2 Red
Compound: Anthocyanin
As chlorophyll is broken down, anthocyanins are produced. These produce leaves of red, purple and pink that are displayed on trees such as maples, oaks and sumacs.

3 Orange
Compound: Carotene
Sugar maple trees turn a vibrant orange colour in autumn due to the increased concentration of carotene pigment. While chlorophyll diminishes with reduced light, carotene doesn't. This increases its concentration, and when combined with anthocyanin, different shades of orange are made.

4 Yellow
Compound: Xanthophyll
This pigment is in a group called carotenoids, along with carotene. However, this pigment contains more oxygen and produces a light-yellow colour. Some of the trees that display xanthophyll in autumn include birches, beeches and aspens.

5 Green
Compound: Chlorophyll
Evergreen plants continue to photosynthesise during autumn. They retain their green colour year-round by continually producing chlorophyll – the green pigment that absorbs sunlight.

UNDERSTANDING BIOLOGY

The grass is greener
The same molecules that feed grass also give it its colour

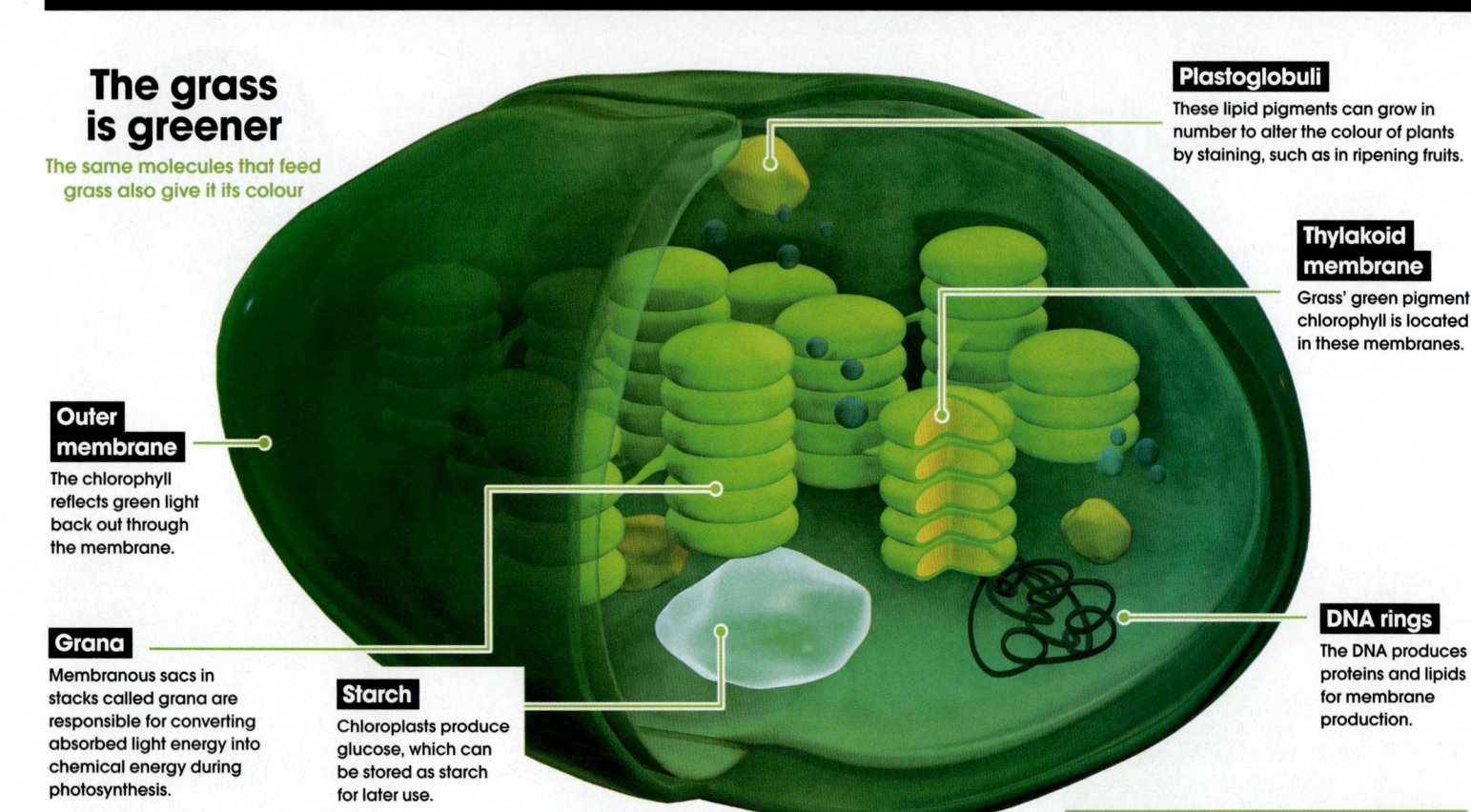

Plastoglobuli
These lipid pigments can grow in number to alter the colour of plants by staining, such as in ripening fruits.

Thylakoid membrane
Grass' green pigment chlorophyll is located in these membranes.

Outer membrane
The chlorophyll reflects green light back out through the membrane.

Grana
Membranous sacs in stacks called grana are responsible for converting absorbed light energy into chemical energy during photosynthesis.

Starch
Chloroplasts produce glucose, which can be stored as starch for later use.

DNA rings
The DNA produces proteins and lipids for membrane production.

Why is grass green?

The same molecules that feed grass also give it its colour

Like many of the plants on Earth, grass is mostly green. This is because there are millions of cells called chloroplasts in every blade of grass. Within these cells is a pigment called chlorophyll. We see grass as green because it is one of the few visible wavelengths from the Sun that chlorophyll doesn't absorb. Instead, after sunlight reaches grass, the green wavelengths are reflected off the plant. So when they reach our eyes, we see grass as green.

But the main role of these cells isn't to look pretty. Grass gets its green colour as a consequence of one of the plant's most vital processes – photosynthesis. Within chloroplasts, the green-giving pigments absorb large amounts of light so that they can use its energy to produce glucose for food. Red and blue wavelengths are prioritised by the plants, as they provide the most energy for photosynthesis. The green wavelengths are reflected while the red and blue are absorbed. While green is the most important colour for us in identifying healthy grass, it's the least important for the plant itself.

Not all grass is the same variety

Losing its colour

At different times of year and in different weather conditions, grass can appear different shades of green. Grass grows at different rates based on its location, type or the weather. In spring, it is likely to grow most rapidly, producing new cells such as chloroplasts. The higher the concentration of chloroplasts within a blade of grass, the more green pigment there will be too. When grass turns brown or yellow it might still be alive, but very low on chlorophyll. If you notice that the bottom of the blades of grass are losing their green colour, this is often due to the amount of water the grass is getting. When grass becomes dry, the lower parts of the blades will begin to turn brown, but if the opposite is true and you are overwatering the grass, it may appear yellow. This is because if the soil is waterlogged, oxygen can't get to the plant and the grass becomes oxygen-deprived.

Green is the colour of healthy grass

PLANTS & FUNGI

Why nettles sting

The science behind the sting of these common countryside plants

On first impression, stinging nettles (Urtica dioica) aren't a particularly threatening-looking plant. However, if you've ever brushed against their leaves, you'll know all too well the pain they can inflict.

What gives nettles their stinging power is a cocktail of chemicals held within tiny hair-like structures along the plant's leaves. When these structures are broken they release their chemical contents, which penetrate the skin and cause several painful symptoms such as itching, swelling and redness. The initial skin irritation felt when you are stung by a nettle is caused by histamine, which is also responsible for causing other allergic reactions.

Like many other plants, nettles have evolved their stinging abilities as a defence mechanism against herbivorous animals such as deer and rabbits. Nettle venom isn't strong enough to seriously harm a hungry herbivore – it merely teaches them a lesson in avoiding nettles.

Commonly found around hedgerows, woodlands and fields, stinging nettles are much more than vicious plants, as they also play an important role in their ecosystem. Butterflies, for example, use nettles as a place to lay their eggs, and ladybirds and aphids use nettles for shelter and as a source of nutritious food.

Chemical cocktail
How nettles deliver their irritating venom

1 Breakaway tip
Across the nettle leaves are spiny hairs called trichomes. The bulbous tip of the trichome breaks off when it's brushed against, revealing the needle-like tube below.

2 Injection
The trichome's inner needles are filled with a stinging cocktail of chemicals including formic acid, histamine, serotonin and acetylcholine.

3 Itching
An itching sensation and a burning rash can occur following a sting, lasting up to 12 hours.

4 Chemical factory
At the base of the trichome are specialised cells that secrete the stinging chemicals into the trichome needle.

Do dock leaves really work?

It's long been believed that grabbing a dock leaf and rubbing it on a wound after a nettle sting will help ease the pain. Dock plants, also known as bitter dock (Rumex obtusifolius), are said to produce a sap that contains an antihistamine to combat the effects of nettle stings. However, there's no evidence of such antihistamines or healing chemical properties. The placebo effect has also been cited as a possible explanation for why some people believe dock leaves help treat the stinging sensation and lower the perception of pain.

One way that the dock leaf sap can offer some relief is through evaporative cooling. As the dock sap evaporates over the sting, the skin beneath can experience a cooling sensation that relieves some of the burning irritation on the skin.

Dock leaves have large, flat leaves and grow in fields and woodlands

A view of a nettle leaf and its many trichomes under a scanning electron microscope

UNDERSTANDING BIOLOGY

Searching for heat

Chilli peppers are comprised of many parts, some much more pungent than others

Placenta
Also called the pith, the white placenta can harbour around 90 per cent of the chilli pepper's capsaicin content.

Capsaicin glands
Embedded in the placenta and situated around the seeds, capsaicin glands produce the compound that gives the pepper its heat.

Seeds
Often incorrectly considered the hottest component, seeds are not particularly dense in capsaicin but are near to the capsaicin-rich glands.

Mesocarp
The centre of the fleshy wall surrounding the internal body of the fruit, the mesocarp contains much of the pepper's water content.

Apex
The least spicy edible part of the pepper, the apex of the fruit lies at its tip, where little capsaicin is found.

The Carolina reaper is the world's hottest pepper at 2.2 million Scoville heat units

Milk contains a protein called casein that can carry away capsaicin, making it much more effective at combating spice than water

What makes chilli peppers spicy?

Meet the tasty chemical compounds that pull sweat from foreheads, force tears from eyes and make tongues feel like they're on fire

Spices have been part of human diets for thousands of years, and chilli peppers are now staple ingredients in many parts of the world. Chilli peppers are fruits and are part of the nightshade family, which contains members such as tomatoes, avocados and potatoes. Within this family lies the genus Capsicum that hosts peppers, a group of related species that are carefully cultivated to produce fruits with a spectrum of flavour and heat. From the mild bell pepper and the lively jalapeño to the unrelenting brutality of the Trinidad Moruga scorpion and Carolina reaper, peppers come in all ranges of spiciness. The heat, or pungency, of a chilli pepper is a product of its DNA, the environment it's grown in and its ripeness. DNA can be changed by cross-breeding two species, followed by selective breeding of the offspring where the progeny with the desired traits are bred further. The stress imposed on the pepper, such as the temperature it's grown in and the amount of water available to it, can likewise affect its pungency. Peppers also become spicier as they ripen from green to red.

While humans have learnt to manipulate and enrich pepper qualities, their spicy trait first evolved naturally as a defence mechanism. Rodents and mammals are equipped with receptors in their mouths that recognise the compound capsaicin – and other compounds collectively known as capsaicinoids – produced within chilli peppers. Capsaicin surrounds the seeds, which are needed for the plant to germinate its progeny, and the compound triggers the sensation of burning when consumed. This helps dissuade rodents and mammals from eating too much, with the latter being particularly threatening to the plant as they can grind and destroy the seeds as they eat the fruit. But birds, which do not grind the seeds and instead are useful helpers in dispersing them, are not sensitive to capsaicin and so are not discouraged from feeding. Capsaicin also wards off microorganisms from invading and decomposing the plant matter from the inside.

If capsaicin production and its spicy heat have evolved to deter hungry mammals from munching through peppers, then why do many humans ravenously ingest them? Spice serves a practical role in food by helping to prevent

PLANTS & FUNGI

WHAT MAKES CHILLI PEPPERS SPICY?

Feel the burn

The presence of capsaicin provokes a reaction from our brain as if we've come into contact with a high-temperature food

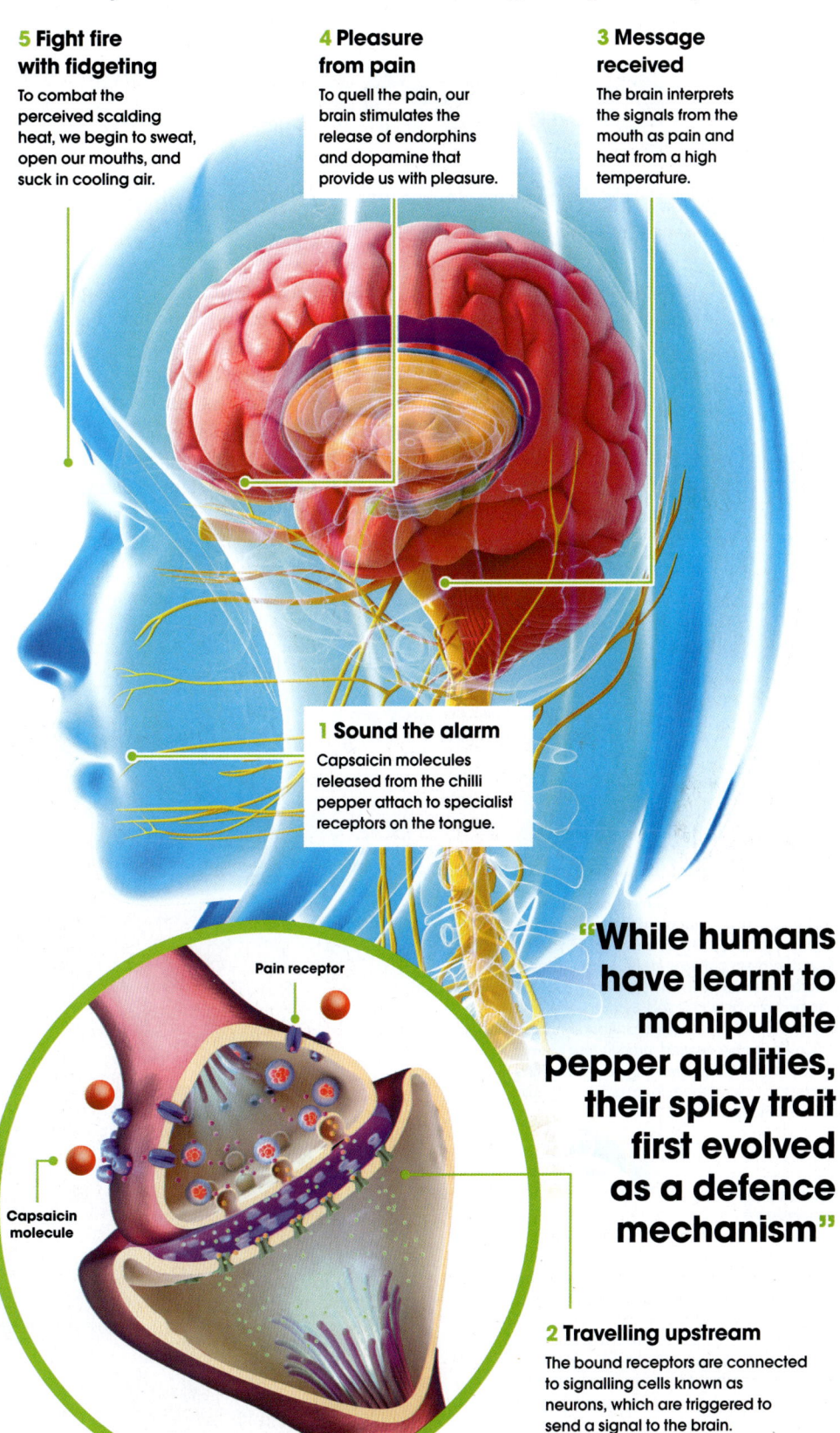

The Scoville scale

In 1912, Wilbur Scoville created the Scoville Organoleptic Test as a means of measuring the amount of heat in a pepper. The method he developed was part quantitative and part subjective. Scoville gathered taste testers and asked them to taste a pepper's capsaicin content – dissolved in an alcohol solution – mixed with sugar water. The capsaicin solution would be continually diluted with sugar water until heat could no longer be detected, which provided Scoville heat units (SHU). A pepper measuring 5,000 SHU would need to be diluted by 5,000 before no heat could be detected. Jalapeños measure between 3,000 and 10,000 SHU, whereas pure capsaicin measures 16,000,000.

Scoville was a chemist who invented the scale measuring pepper pungency

5 Fight fire with fidgeting
To combat the perceived scalding heat, we begin to sweat, open our mouths, and suck in cooling air.

4 Pleasure from pain
To quell the pain, our brain stimulates the release of endorphins and dopamine that provide us with pleasure.

3 Message received
The brain interprets the signals from the mouth as pain and heat from a high temperature.

1 Sound the alarm
Capsaicin molecules released from the chilli pepper attach to specialist receptors on the tongue.

2 Travelling upstream
The bound receptors are connected to signalling cells known as neurons, which are triggered to send a signal to the brain.

Pain receptor

Capsaicin molecule

> "While humans have learnt to manipulate pepper qualities, their spicy trait first evolved as a defence mechanism"

spoilage, but many of us who enjoy spice also like the challenging heat. Capsaicin can trigger the release of endorphins that provide pleasure as a means to combat the pain.

However, the pleasure of eating spice is not for all of us. The endorphin release does not happen for everyone equally, and for those who are sensitive or who overindulge, the pleasure can quickly recede while the pain intensifies. Unfortunately, the damage of too much spice doesn't end in our mouths, as capsaicin can bind to receptors in our stomach and intestines. This can trigger diarrhoea as our body acts to protect itself by rapidly funnelling the irritant compound through the gut as quickly as possible. And the receptors that interpret capsaicin as heat at the beginning of our gastrointestinal tract are also there at the end, which is why we can also feel burning as we eliminate stool. Finally, as capsaicin is an irritant that also affects our outer organs, the spiciest peppers can elicit burning sensations when they contact our skin and eyes.

UNDERSTANDING BIOLOGY

How to eat poisonous plants

Here's how toxic compounds are removed from some of our foods

There are plenty of species of plants around the world that contain substances that could either kill you or cause you harm. However, there are also some species where humankind has figured out ways to bypass or remove them, making the plants safe to eat.

For thousands of years, Aboriginal groups in northern Australia have been eating plant seeds that are so toxic that, when ingested, just two are strong enough to kill a canine. These potent plants are called cycads and can be found in tropical and subtropical regions around the world. The seeds of these plants are filled with a toxin called cycasin, which can wreak havoc on the human body. Once inside the body's digestive system, intestinal bacteria alter the cycasin, releasing a toxic metabolite called methylazoxymethanol. This metabolite enters the hepatic portal vein, the vein that supplies blood from the spleen and intestines to the liver, and starts damaging liver cells. The damage can lead to hepatic necrosis (death of liver cells), gastrointestinal irritation, liver failure and death.

Over time, Aboriginal peoples developed several ways to avoid the poisonous power of these seeds to access their high nutritional value. One of the most successful methods was

Poison pantry

The raw foods you didn't know were filled with toxins

Kidney beans

Dried beans, such as kidney beans and soya beans, are toxic. Uncooked kidney beans contain something called phytohaemagglutinin (PHA), a type of protein called a lectin. Lectins play many different roles in the body, but some, such as PHA, can be toxic at high levels. Symptoms of PHA poisoning include violent diarrhoea and vomiting, which ease after around three to four hours. Simply soaking the raw beans for at least five hours and boiling them for at least 30 minutes destroys the toxin.

Cherries

Sitting in the centres of many fruits are toxins. Stoned fruits, also known as drupes, such as cherries and peaches contain a central seed or pit surrounded by fleshy fruit. Within the seed is amygdalin. Similar to almonds, when ingested it's converted by the body into cyanide. Around 200 raw cherry seeds contain around 117 milligrams of cyanide – a lethal dosage of cyanide ranges from 0.5 to three milligrams per kilogram of body weight.

PLANTS & FUNGI

HOW TO EAT POISONOUS PLANTS

prolonged leaching, whereby the seeds were harvested, dried and pounded before being left to soak in either streams or human-made waterholes for between three and seven days.

This preparation broke down the cell membranes of the seeds and allowed the toxin to leach out of them. Aboriginals also found that ageing the seeds underground for several months achieved similar results. Like cycads, there are many other seeds, nuts and beans that need to be treated before they are safe to eat. For example, almonds are a popular snack or ingredient around the world, with around 1.56 million tonnes of them consumed globally each year. However, almonds have to undergo a pasteurisation process to make them safe to eat – they can't be eaten straight from the tree.

In their raw state, almonds contain a chemical called amygdalin, which is broken down by the body into the toxin cyanide. Once ingested, cyanide interferes with the ability of human cells to obtain energy from oxygen, causing the cells to die. Eating between six and ten raw untreated almonds can cause serious poisoning in the average adult, and consuming around 50 or more can cause death. To avoid harming us, almonds undergo a process of pasteurisation, similar to the treatment of milk. Batches of almonds are either heated via steam, roasted, boiled or even chemically treated to break down the amygdalin before the 'raw' seeds hit the supermarket shelves.

It goes without saying that you should never try to eat a food that is poisonous.

Cycad seeds are highly toxic, yet they have been safely eaten for thousands of years

Almonds being treated for consumption

Rhubarb
Similarly to potatoes, rhubarb also has toxic leaves. Although the stems are delicious, their leaves are packed with oxalic acid – around 0.5 grams per 100 grams of leaves. Some signs of oxalic acid poisoning include vomiting, abdominal pain, convulsion and even red-coloured urine. The acid is fatal when between 15 and 30 grams is ingested, which would be between three and six kilograms of rhubarb leaves.

Potatoes
Don't worry, you've not been eating poisonous potatoes this whole time. Not unless you've been eating the leaves, stem or sprouts of the plant that is. The green parts of the potato plant are packed with toxins called solanine and chaconine. This toxic duo can cause intestinal issues such as diarrhoea, vomiting and abdominal pain when ingested. The white, edible portion of the plant has very low concentrations of the toxins, and cooking potatoes eliminates the solanine and chaconine.

Black liquorice
This common sweet treat contains a compound called glycyrrhizin. Although harmless in moderation, when consumed in excess, liquorice root and sweets that contain liquorice root can be deadly. Glycyrrhizin lowers levels of potassium in the body, which can lead to high blood pressure and abnormal heart rhythms. Although cases of fatality are extremely rare, eating around 57 grams of black liquorice a day for two weeks can cause heart problems.

A cycad seed is around five per cent proteins

UNDERSTANDING BIOLOGY

Toxic Mushrooms
Why some fungi are so poisonous and how to spot them

Did you know?
Squirrels and rabbits can eat death caps unharmed

PLANTS & FUNGI

TOXIC MUSHROOMS

Earth's forests are littered with the weird and wonderful umbrellas of mushrooms – some edible and some deadly. Mushrooms are classified as the reproductive structures of some, but not all, species of fungi. Similarly to how fruits bear the seeds of some plants, mushrooms are filled with millions of spores, held in their gills. These spores ride the wind or hijack a lift on a passing animal and travel to their new home to begin another mushroom population.

There are around 14,000 different mushroom species within the 3.8 million species of fungi on Earth. Of those thousands, around one per cent are poisonous and a handful of them are deadly enough to kill a human. Through ingestion, and in rare cases through touch, mushrooms can deliver a collection of toxins that wreak havoc on the human body. This can include internal organs failing, gastrointestinal issues and can even cause parts of the brain to shrink.

The most popular theory behind why some mushrooms have evolved to be so poisonous is for protection from being eaten by certain species, including humans. One of the leading causes of mushroom poisoning is a lack of knowledge in identifying fungi and misidentifying harmful mushrooms for edible alternatives. Here are a few of the worst offenders.

A poison fire coral emerging through a mossy forest floor

Poison fire coral
Podostroma cornu-damae

Poison fire coral looks like it belongs under the sea rather than in the woods, with its reddish finger-like protrusions emerging from the soil. Nevertheless, this potent fungus can be found on tree roots across Asia, including Japan, Korea and Java. Unlike many other deadly mushrooms that are poisonous when eaten, the toxins in poison fire coral can be absorbed straight into the skin. Touching the fungus can cause inflammation and dermatitis, a skin irritation. When eaten, toxins in these mushrooms called trichothecene mycotoxins can cause the skin to peel, hair loss and even cause shrinkage of the cerebellum – the portion of the brain involved in movement and speech.

Unfortunately, these mushrooms have mysteriously found their way from Asia to Queensland, Australia. Although scientists are unsure as to how and when the mushroom migrated, some researchers have suggested its spores may have made the journey down under thousands of years ago.

A singular death cap growing through autumn leaves

Death Cap
Amanita phalloides

The most poisonous mushroom on Earth is the death cap, accounting for 90 per cent of all deaths caused by fungi. Death caps are common throughout forests in the UK, Ireland and Europe and often grow near oak and beech trees. However, they can be found on almost every continent.

These deadly mushrooms don't look particularly threatening or too dissimilar from grocery-variety mushrooms. However, they contain a highly potent toxin that is strong enough to kill a person. Eating just half of the mushroom cap can be fatal. The initial symptoms of death cap poisoning include diarrhoea and vomiting, causing severe dehydration. The amatoxins – a group of toxins found in other mushroom species – within the death cap affect the function of the human body's liver cells, called hepatic cells. The toxins are eliminated from the body via the kidneys and urine, but without swift medical intervention the hepatic cells will die and the liver will fail around two to three days after ingestion.

Deadly webcap
Cortinarius rubellus

Webcaps are among the largest known genera of fungi on Earth. Around 2,000 to 3,000 species have been identified. However, this deadly webcap is rarely found outside of northern Europe. Symptoms of eating a deadly webcap include vomiting, diarrhoea, headaches and ultimately kidney failure. These mushrooms have a potent toxin called orellanine. When ingested, orellanine produces oxygen radicals that damage kidney cells and can cause renal failure. The last known case of webcap poisoning in the UK was in 1979. The species has been confused with the edible chanterelle (Cantharellus cibarius) mushroom.

A brown deadly webcap growing in a forest in Croatia

Fool's conecap
Conocybe filaris

This is a widely distributed lawn mushroom that can be found in Europe, Asia and North America, particularly in the Pacific Northwest region of the US. Amatoxins are this mushroom's poison of choice. The amatoxins found in a fool's conecap are made up of nine different types of toxins that together prevent vital proteins from being synthesised in liver and kidney cells, which can be fatal. Those that ingest it develop gastrointestinal symptoms within 24 hours. Symptoms can be overlooked as food poisoning or stomach flu. This mushroom is often mistaken for hallucinogenic mushrooms such as psilocybin mushrooms. However, unlike psilocybin mushrooms, their cone-shaped caps will eventually open out into an umbrella.

A group of fool's conecap mushrooms, found in Lower Rhine, Germany

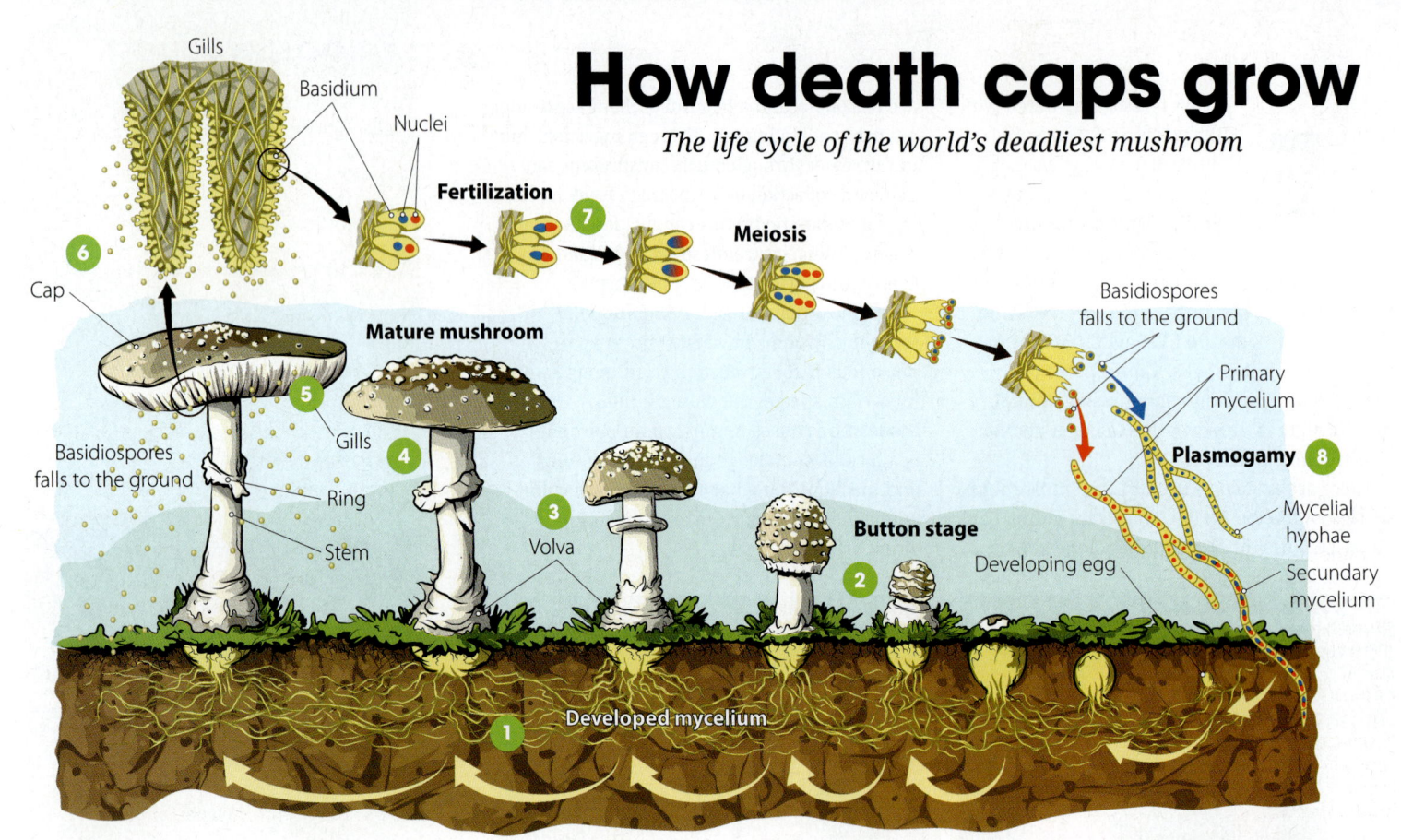

UNDERSTANDING BIOLOGY

How death caps grow
The life cycle of the world's deadliest mushroom

1 Mycelium
These are root-like fungal structures from which mushrooms grow. They break down nutrients in the soil for food.

2 Button stage
The first mushroom structure to appear above ground. It's odourless, unlike its sweet-smelling adult form.

3 Volva
The universal veil – a layer of tissue that once surrounded the baby mushroom – forms a bag at the base of the adult mushroom called the volva.

4 Mature mushroom
A mature death cap mushroom will measure around 15 centimetres tall and have a domed white cap with a white stem.

5 Gills
In the gills of the mushroom are structures called basidia that bear reproductive spores.

6 Spores
Billions of spores are released from the gills, carrying the genetic information of the mushroom away.

7 Germination
Just like reproductive cells in humans, called gametes, spore cells undergo meiosis, whereby they divide and multiply and form long structures called primary mycelium.

8 Plasmogamy
At this point, two primary mycelium fuse together to make one final mycelium that develops and grows into a ingle mushroom.

"Of those thousands, around one per cent are poisonous and a handful of them are deadly enough to kill a human"

Autumn skullcap Galerina marginata

Also known as deadly galerina or funeral bell, these brown mushrooms pack a punch in the poison department. Similar to the death cap and destroying angel, autumn skullcaps are filled with amatoxins that target the liver and cause hepatic cell failure. Although the concentrations of these toxins are lower than other deadly species, they can still cause vomiting and ultimately liver failure around 72 hours after ingestion. Amatoxins are also heat-stable, so cooking these mushrooms doesn't remove or break the toxins down.

These fungi like to grow on decaying hardwood and softwood, which appear to contribute to the number of different amatoxins within the mushroom along with its genetic make-up and other environmental factors.

Poisonous Galerina marginata growing in a floodplain forest

PLANTS & FUNGI

TOXIC MUSHROOMS

A destroying angel growing in a forest in Minnesota

Destroying angel Amanita virosa

As angelic as this mushroom may appear, its poisonous biology makes it devilishly dangerous. Closely related to the death cap, the destroying angel, also known as the death angel, contains a cell necrosis-inducing amatoxin. Research has found that these toxins can induce apoptosis, the process of programmed cell death. Like other mushrooms containing amatoxins, poisoning symptoms begin as gastrointestinal irritation and later develop into more serious damage to the kidney and liver. Unlike its more toxic cousin the death cap, ingesting destroying angels isn't always a death sentence, and the human liver can completely recover in some cases. These white mushrooms can be found in mixed oak and hardwood forests across Europe and the US.

Staying safe around mushrooms

David Winnard is a foraging expert and founder of Discover the Wild, a UK-based natural history company that offers foraging and species identification services. Winnard is a renowned naturalist in the UK and has worked with many conservation organisations and local councils

How do you spot a poisonous mushroom? What to look out for?
There's no clear way of saying if it has this, it's edible; if it has that, it's poisonous. It's not as simple as that. What we have are families which are generally considered edible which have poisonous members. Similarly, you get very poisonous family groups that happen to have some edible ones. Learn edibles that have no poisonous lookalikes. In parts of Europe, this is what's done. Learn different edible species, stick to those and don't deviate. For example, porcini mushrooms don't have anything that looks like them that is going to kill you. With foraging, identify the edible ones and ignore everything else to eliminate all the poisonous ones.

If you see a death cap under stress or in dry conditions it can look similar to blusher [Amanita rubescens], and they grow in the same habitat. If you're going along with a basket and not really thinking about what you're doing, a death cap can end up in your basket very easily. I tell people just don't eat anything from the Amanita family, even if it's edible. There's no mushroom I've tried that is worth dying for.

Why is the death cap so dangerous?
Death cap is a particularly worrying one because it has no mechanism for people to stop eating it. With a lot of mushrooms, they're bitter or foul-tasting or they smell bad. Whatever it is, your body reacts and says spit it out. Death cap doesn't have that. It smells sickly sweet – like honey in some cases – you can almost smell them before you see them.

What advice do you have for anyone thinking of foraging wild mushrooms in the UK?
Understand your limits about what you do and don't know. Be brave enough to say you don't know. If you are 99.999 per cent sure, do not eat it. You need to be 100 per cent certain. You also need to have the discipline to take the whole specimen home and confirm what you think it is with good reputable field guides, not apps. I've field-tested quite a few [mushroom identification] apps, and I think this is where more recent poisonings are probably coming from because people are relying on pointing their phone at a mushroom and seeing what it is. These apps are so far away from being reliable for mushrooms. For indoor house plants they're great, because plants are very formulaic [in their appearance] – mushrooms are not.

UNDERSTANDING BIOLOGY

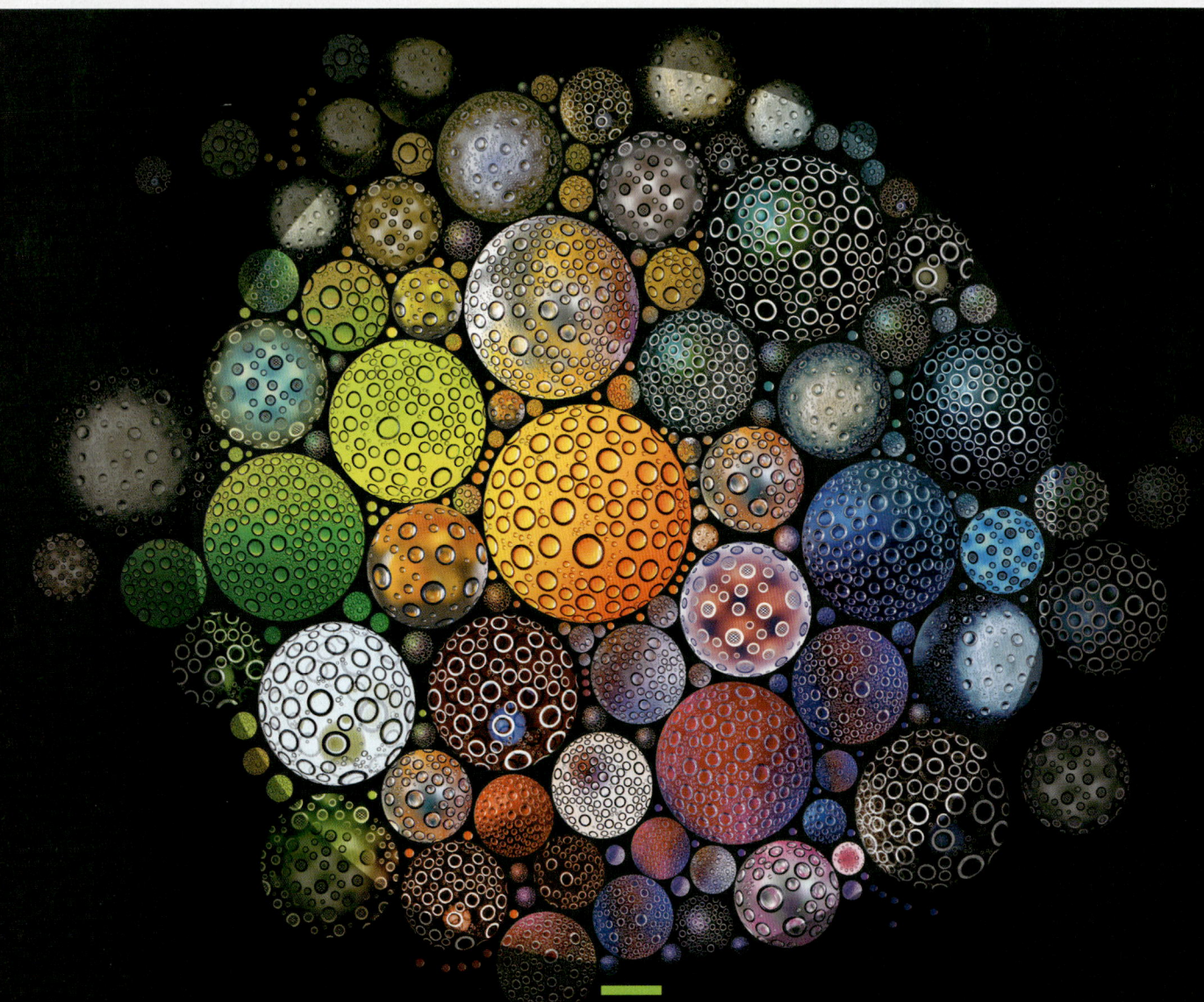

The weird world of mould

The strangest, deadliest, most disgusting and useful moulds found on Earth

The world's mould species belong to the fungi kingdom. However, the two are distinctly different. Most fungi can be either multicellular or unicellular organisms, feasting on organic matter such as the carbon and glucose content found in decaying leaves. Moulds, on the other hand, are only ever multicellular organisms, which tend to extract nutrition from many different sources.

There are more than 100,000 different species of moulds on Earth, many of which can be found in your home. As part of their reproductive cycle, moulds release their version of seeds, called spores. Moulds release hundreds of thousands of these, which measure between 1.0 and 40 micrometres, into the air – for comparison, the diameter of a human hair is 70 micrometres. Some of the spores will find a new home and begin to grow under the correct conditions.

Moisture is a key ingredient for a mould's development. Damp walls, leaky roofs and spoiling fruit are all excellent sites for a mould spore to land. Once the spore has landed it develops into fine threads called hyphae. These intertwine and form a network called a mycelium. Unlike plants, mould mycelium cannot produce its energy from photosynthesis, and so draws nutrition from its surroundings by releasing digestive enzymes to extract carbon and sugars and convert them into energy.

PLANTS & FUNGI

THE WEIRD WORLD OF MOULD

Bread binger
Rhizopus stolonifer is a common mould found on bread

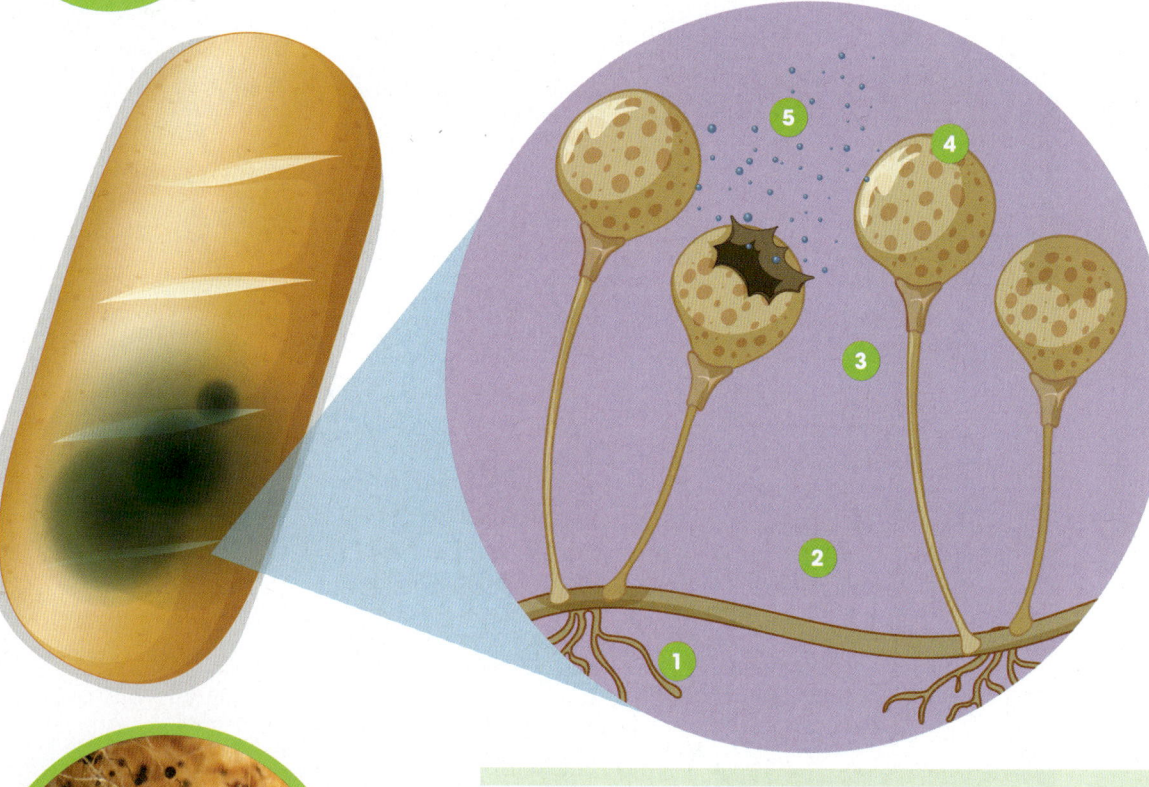

Did you know?
The largest single fungus, Armillaria ostoyae, covers around 3.7 square miles

A close-up view of a bread mould's sporangia

1 Rhizoid
The root-like anchor of the mould, this releases digestive enzymes into the substance below.

2 Stolon
The horizontal network of hyphae from which spore-carrying bodies emerge.

3 Sporangiophores
The stalks that hold the spore nurseries, called sporangia.

4 Sporangia
At the apex of the sporangiophores are the spherical 'fruits' that house the mould spores.

5 Spores
The spores are the mould's reproductive cells. They float through the air, land on more bread to multiply and form new mould.

The miracle fungus

If you've ever had a serious bacterial infection, you're likely to have been treated with a little white pill made from mould. Discovered by chance in 1928, Penicillium chrysogenum can fight invading bacterial forces and prevent them from regrowing. While studying Staphylococcus bacteria, Dr Alexander Fleming discovered that after some time away, mould had grown in his study Petri dishes that were filled with bacteria. He witnessed that no bacteria were growing around the mould, soon concluding that it must produce a chemical defence to ward off the bacteria. He called it penicillin. Scientists now know that what Fleming observed was penicillin's ability to break down the membrane of bacteria, allowing it to be tackled more easily by the body's immune system. More than ten years after his discovery, a group of scientists uncovered Fleming's work – which had been shelved by Fleming – and produced a pure penicillin drug to fight bacterial infections.

An artist's illustration of Penicillium chrysogenum

4 Moulds that make you sick

1 Cladosporium
A common mould found on plants, in soil and on many household surfaces, the spores of this mould can trigger allergies and asthma symptoms after long-term exposure.

2 Fusarium
Mycotoxins released by this mould can affect the performance of its host's immune system and negatively impact gastrointestinal cell lining if ingested.

3 Aspergillus
A common mould that doesn't affect the majority of people. However, its spores can cause allergic reactions and lung infections in people with weakened immune systems.

4 Penicillium
Although it's used to make medicines, this mould can also trigger asthma symptoms among sufferers and surface infections such as keratitis, an inflammation of the eyes' corneae.

127

UNDERSTANDING BIOLOGY

The dark side

Many moulds are harmless to humans – some are even used to enrich the flavour of some cheeses, such as Stilton and Roquefort. However, the inhalation or ingestion of some mould spores can be hazardous. Black mould (Stachybotrys chartarum) is one of the worst that can find its way into your home. Needing only 12 to 24 hours to grow, it's prolific under the right conditions. Once it's established itself, the spores can trigger the body's immune responses, resulting in symptoms such as sneezing and a runny nose. However, in extreme cases, long-term exposure to black mould has been linked with a potentially life-threatening condition in infants called acute pulmonary haemorrhage, though further research is required.

Black mould, like many other mould and fungi species, can release mycotoxins that can disrupt or damage different systems in the body. For example, Byssochlamys is often found on rotting apples and releases a mycotoxin called patulin. When consumed, these toxins can cause vomiting and gastrointestinal issues in humans, but can lead to kidney or liver damage in some animals. To avoid the hazardous effects of mould exposure, ensure that your home stays as dry as possible to prevent growth and seal off areas that are infested until they are cleaned. Toss any food that has begun to grow a mould colony and don't just scrape it off the top... mould mycelium can burrow much deeper into food than you think.

It can take less than a day for black mould to appear on damp walls

Symptoms of black mould exposure

1 Sneezing
This is the body's way of removing trapped spores from the mucus in the nose.

2 Coughing
Spores trapped in the mucus lining of the lungs can be removed through coughing.

3 Red eye
Spores and mycotoxins can irritate the eyes, causing them to appear bloodshot.

4 Postnasal drip
Part of the body's immune response to spore infiltration is overproducing mucus, which can drip to the back of the throat.

5 Wheezing
Symptoms of existing respiratory conditions such as asthma can be worsened, including shortness of breath.

Slimy mystery
These unusual moulds have strange abilities

Slowly crawling along the forest floor are a diverse group of organisms we've come to know as slime mould. These unusual life forms were once thought to be a type of fungus. However, scientists have discovered that they belong to another group called Protozoa. This group of organisms includes free-moving single-celled organisms such as amoebae. Once slime mould has matured, it will cease to move and bear sporangia 'fruits', similar to moulds. These fruits will then release spores that will go on to grow into new slime moulds. However, unlike mould and other fungi, slime moulds don't have penetrating hyphae to dig into wood and extract nutrients. Instead their amoeba-like movements allow them to freely crawl their way along the surfaces of plants and trees, engulfing bacteria along the way. One slime mould, Physarum polycephalum – nicknamed 'the blob' – has piqued the interest of scientists. Despite not having a brain or nerve cells, scientists have discovered that this strange slime remembers things and can solve mazes. It's not yet fully understood how the slime mould retains this kind of information.

Physarum polycephalum looks a bit like coral

PLANTS & FUNGI

THE WEIRD WORLD OF MOULD

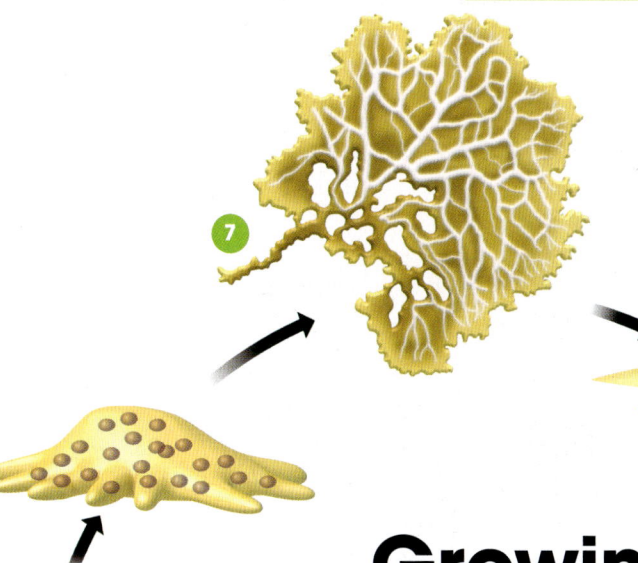

Growing slime mould

How a bunch of cells come together to form this odd organism

Did you know? Slime moulds have been recorded travelling at speeds of up to 1.35 millimetres per second

1 Sporangia emerge
The spore-bearing structures begin to sprout from a mature mass of slime mould, called a plasmodium.

2 Spore release
Once a sporangium is fully formed and filled with spores it ruptures, releasing them onto nearby decaying wood.

3 Germination
From these spores, one or two new cells emerge – either an amoeboid cell under dry conditions or a flagellating cell under wet conditions.

4 Conversion
If necessary, either the amoeboid cells or flagellated cells can transform into the other to suit a change in their environment.

5 Feeding
During the amoeba stage of their lives, slime moulds will feed on bacteria until food becomes scarce.

6 Plasmogamy
When their food runs out and the amoebae begin to starve, they fuse together, rapidly multiplying and forming a larger body known as a plasmodium.

7 Maturity
When mature, the multicellular plasmodium will begin to spread out into a netting shape in preparation to grow sporangia and release spores.

Tubulifera arachnoidea slime mould in its plasmodium stage

5 Facts Strange slimes

1 False puffball (Enteridium lycoperdon)
This slime mould looks similar to the fungal giant puffball (Calvatia gigantea) during its reproductive stage of life. False puffballs are often found hanging from dead trees.

2 Carnival candy (Arcyria denudata)
The vibrant red 'fruits' of this slime mould resemble sponge-like loofahs. They release their spores onto nearby decaying wood for their amoeba hatchlings to feast on the bacteria within.

3 Red raspberry (Tubifera ferruginosa)
The sporangia of this slime mould are so tightly packed together that they resemble a raspberry. The vibrant sporangia eventually turn a purplish-brown before releasing spores.

4 Coral slime (Ceratiomyxa fruticulosa)
Found in forests around the world, this slime mould transforms its body into icicle-like stalks that bear the mould's spores.

5 Comatricha nigra
When this mould, which grows on decayed wood on forest floors, is ready to release its spores, it grows individual lollipop-like sporangia.

UNDERSTANDING
Biology

Future PLC Quay House, The Ambury, Bath, BA1 1UA

Editorial
Editor **April Madden**
Senior Art Editor **Andy Downes**
Head of Art & Design **Greg Whitaker**
Editorial Director **Jon White**
Managing Director **Grainne McKenna**

How It Works Editorial
Editor **Ben Biggs**
Group Editor-in-Chief **Tim Williamson**
Senior Art Editor **Duncan Crook**
Production Editor **Nikole Robinson**
Senior Staff Writer **Scott Dutfield**
Staff Writer **Ailsa Harvey**

Cover images
Getty Images

Photography
All copyrights and trademarks are recognised and respected

Advertising
Media packs are available on request
Commercial Director **Clare Dove**

International
Head of Print Licensing **Rachel Shaw**
licensing@futurenet.com
www.futurecontenthub.com

Circulation
Head of Newstrade **Tim Mathers**

Production
Head of Production **Mark Constance**
Production Project Manager **Matthew Eglinton**
Advertising Production Manager **Joanne Crosby**
Digital Editions Controller **Jason Hudson**
Production Managers **Keely Miller, Nola Cokely, Vivienne Calvert, Fran Twentyman**

Printed in the UK

Distributed by Marketforce, 5 Churchill Place, Canary Wharf, London, E14 5HU www.marketforce.co.uk – For enquiries, please email: mfcommunications@futurenet.com

Understanding Biology First Edition (HIB5837)
© 2024 Future Publishing Limited

We are committed to only using magazine paper which is derived from responsibly managed, certified forestry and chlorine-free manufacture. The paper in this bookazine was sourced and produced from sustainable managed forests, conforming to strict environmental and socioeconomic standards.

All contents © 2024 Future Publishing Limited or published under licence. All rights reserved. No part of this magazine may be used, stored, transmitted or reproduced in any way without the prior written permission of the publisher. Future Publishing Limited (company number 2008885) is registered in England and Wales. Registered office: Quay House, The Ambury, Bath BA1 1UA. All information contained in this publication is for information only and is, as far as we are aware, correct at the time of going to press. Future cannot accept any responsibility for errors or inaccuracies in such information. You are advised to contact manufacturers and retailers directly with regard to the price of products/services referred to in this publication. Apps and websites mentioned in this publication are not under our control. We are not responsible for their contents or any other changes or updates to them. This magazine is fully independent and not affiliated in any way with the companies mentioned herein.

FUTURE Connectors. Creators. Experience Makers.

Future plc is a public company quoted on the London Stock Exchange (symbol: FUTR)
www.futureplc.com

Chief Executive Officer **Jon Steinberg**
Non-Executive Chairman **Richard Huntingford**
Chief Financial and Strategy Officer **Penny Ladkin-Brand**

Tel +44 (0)1225 442 244